GIS *for* Environmental Management

Robert Scally

ESRI Press
REDLANDS, CALIFORNIA

ESRI Press, 380 New York Street, Redlands, California 92373-8100

 First edition 2006
10 09 08 07 06   1 2 3 4 5 6 7 8 9 10

Printed in the United States of America

Library of Congress Cataloging-in-Publication Data
Scally, Robert, 1958–
GIS for environmental management / Robert Scally.—1st ed.
p. cm.
ISBN 1-58948-142-9
1. Environmental management. 2. Geographic information systems. I. Title.
GE300.S26 2006
910'.285—dc22 2006010161

ISBN-13: 978-1-58948-142-8
ISBN-10: 1-58948-142-9

Ask for ESRI Press titles at your local bookstore or order by calling 1-800-447-9778. You can also shop online at www.esri.com/esripress. Outside the United States, contact your local ESRI distributor.

ESRI Press titles are distributed to the trade by the following:

In North America, South America, Asia, and Australia:
Independent Publishers Group (IPG)
Telephone (United States): 1-800-888-4741
Telephone (international): 312-337-0747
E-mail: frontdesk@ipgbook.com

In the United Kingdom, Europe, and the Middle East:
Transatlantic Publishers Group Ltd.
Telephone: 44 20 7373 2515
Fax: 44 20 7244 1018
E-mail: richard@tpgltd.co.uk

Cover design by Jennifer Jennings
Book design and production by Jennifer Jennings
Copyediting by Tiffany Wilkerson
Cartography by Michael Law
Printing coordination by Cliff Crabbe

Contents

Foreword

Earth's environment is a single, vast interdependent system. We cannot make demands on the environment in one part of the world without creating consequences in another. Humanity's effect on these interdependent ecosystems presents difficult challenges for governments, scientists, businesses, and environmentalists in every discipline. Finite natural resources and an ever-increasing population mean responsible and successful environmental management worldwide is no longer a luxury, but rather a necessity.

Protecting and restoring the environment requires better ways of managing our world using informed, responsible, and successful techniques. GIS technology is playing an increasingly crucial role in the delivery of information to decision makers, environmental managers, and the public.

GIS for Environmental Management showcases some of the most innovative GIS projects created by governments, businesses, and individuals. Balancing technology with practical human applications, these case studies give voice to the people using GIS to better manage our environment.

Jack Dangermond
President, ESRI

Preface

Earth's increasingly complex environmental challenges demand increasingly sophisticated solutions. Geographic information systems (GIS) technology is one solution to humanity's need to better manage, protect, and preserve our environment.

Along the U.S.–Mexico border, choking dust threatened the health of residents in Douglas, Arizona, and nearby Agua Prieta, Mexico. GIS identified the sources of the dust and aided in the efforts to solve the problem.

The Missouri Botanical Garden used GIS-based data gathered during twenty-five years of research to help Madagascar expand its ecologically protected areas as part of a campaign to preserve the earth's biodiversity.

In Japan, a university professor created an Internet site that uses GIS to link scientists with others interested in preserving a historic wetland. Another professor in the United States relies on GIS to assess the health of wetlands along the shore of Lake Ontario in upstate New York.

These are but a few of the innovative projects presented in *GIS for Environmental Management*. The projects showcase solutions in various aspects of environmental management, ranging from biodiversity and pollution to more specific subjects such as coastal zone management and habitat change detection.

During the 2005 ESRI International User Conference keynote address, renowned zoologist and conservationist Dr. Jane Goodall said, "Anywhere I see a problem, I see groups of dedicated people trying to right those problems." The theme of the 2005 User Conference, "GIS—Helping Manage Our World," dovetails with this book's message. While the job of managing our environment grows more complex, GIS is helping make the task more manageable.

The case studies in this book are about more than just technology. They also discuss the people using GIS to preserve our environment for future generations. In each of these cases, GIS professionals and scientists devised projects and conducted analyses that would be impossible or impractical without computers and GIS. In several of these projects, solving the technological problems was the easy part. Getting disparate organizations and individuals to work together to create GIS-based solutions that met a shared need or solved a common problem were the projects' true accomplishments.

Acknowledgments

I want to thank Nick Thomas of ESRI's marketing department for conceiving the vision for this book, sorting through more than two hundred potential case studies, and being a great listener and a good friend. I would like to acknowledge two former ESRI Press staff members, R. W. Greene and Christian Harder, for providing support and guidance early in this project.

My editor at ESRI Press, Mike Kataoka, provided his considerable skills and patience in the editing of this book and helped guide it to completion. Thank you, Mike.

Thanks also go to Gary Geyer of Centerville, Indiana, who generously donated his time and finely honed knowledge of the English language. He read one of the final drafts of this book and made edits and suggestions. Gary, you are one of the nation's great teachers; your students are fortunate to have you.

Most importantly I want to thank my wife, Lisa, who is also a great writer and editor, for all of her help, support, and patience. I couldn't have done it without you. Last, but not least, I want to thank my children, Clark and Ava, for their love and support.

I commend all of the people involved in the case studies in this book for their professionalism, depth of knowledge, and cooperation in the making of *GIS for Environmental Management*. Thank you all.

Biodiversity 1

Expanding Madagascar's national parks and protected areas

Madagascar's remote location and unique climate have made it one of the earth's most biologically diverse places.

Data provided by 2004 ESRI Data & Maps.

For botanists and biologists, Madagascar is a fabled lost world alive with numerous unique plants and animals. Eighty percent of the island's twelve thousand to fourteen thousand plant species are endemic, including 850 of 1,000 orchid species. Madagascar is one of the earth's most geographically remote and biologically diverse places and its fourth largest island. It sits in the Indian Ocean about 200 miles (300 kilometers) off the eastern coast of Africa and is about the size of France, its former colonial ruler. A combination of its size, remote location, and varied climate create unique biological environments.

"Madagascar is a living laboratory of evolution," said Dr. George Schatz, biologist and curator at the Missouri Botanical Garden (MBG) in St. Louis, Missouri, and expert on the island's flora.

A new conservation vision

Due to its unparalleled biodiversity, Madagascar is one of the world's top conservation priorities. In 2003, Madagascar's president, Marc Ravalomanana, proposed a major expansion of the nation's national parks and environmentally protected areas. Ravalomanana's conservation plan, known as the Durban Vision, proved fortuitous for MBG researchers. The MBG is a leader in the study of Madagascar's flora and conducted continuous research and numerous conservation projects on the island for more than twenty-five years. Much of the MBG's data is managed within an ArcGIS system, which helped the scientific analysis and created informative maps.

The Malagasy government asked the MBG to recommend plant life to include in the country's expanded protected areas. The MBG used GIS to tap into its vast Malagasy botanical research database to rapidly assemble a comprehensive and authoritative report.

With a permanent research presence in Madagascar, MBG has helped train several Malagasy scientists and, more recently, GIS professionals. When President Ravalomanana conceived of the Durban Vision, MBG researchers happened to be assessing priority sites for plant conservation. That multiphase project, the Assessment of Priority Areas for Plant Conservation (APAPC), was underwritten by the Critical Ecosystem Partnership Fund and is ongoing. The APAPC project fits well with the objectives of the Durban Vision project.

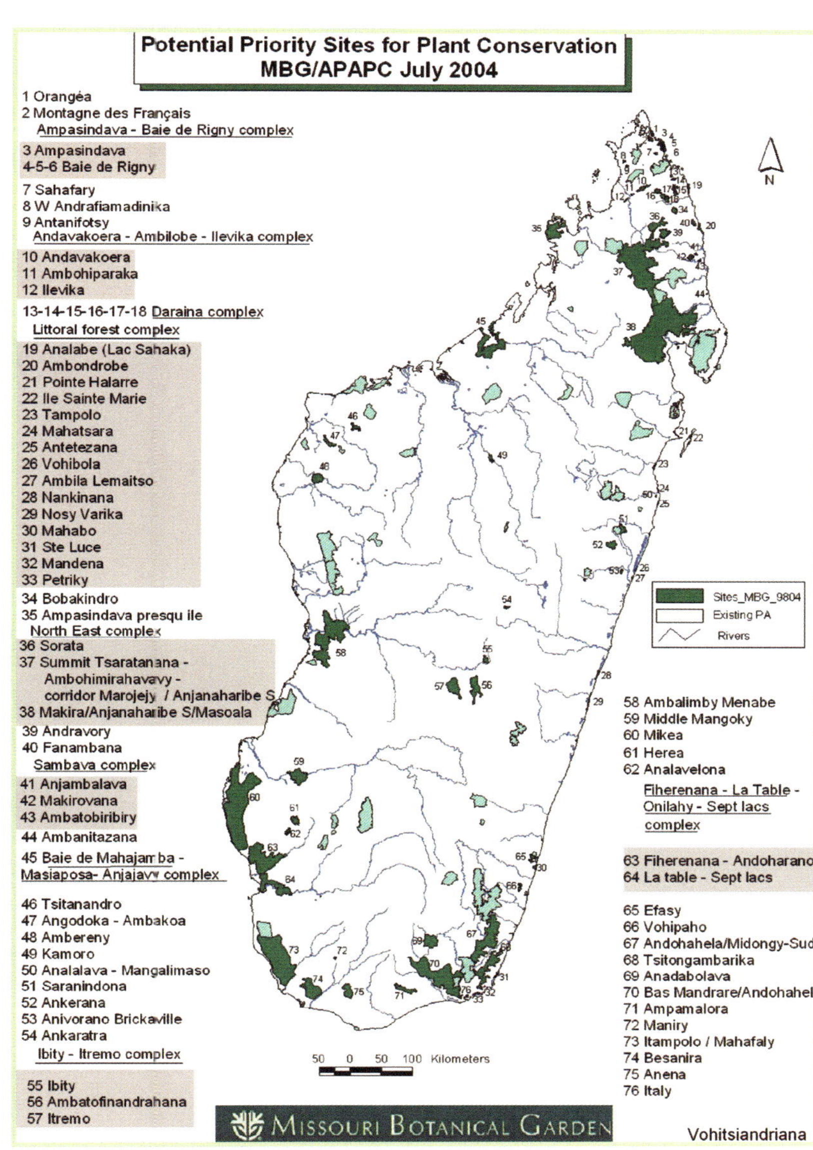

Map by Tantely Raminosoa, Missouri Botanical Garden.

Figure 1.1 Seventy-six sites of plant conservation importance proposed in a two-day workshop at the Missouri Botanical Garden (MBG). Participants included MBG botantists who have worked in Madagascar for more than fifteen years, as well as several Malagasy biologists.

Photo by David Rabehevitra, Missouri Botanical Garden.

Figure 1.2 The littoral forests of eastern Madagascar, such as Nankinana, located on the island's east coast forest north of the city of Nosy Varika, have shrunk rapidly since humans arrived about two thousand years ago. The few remaining patches of littoral forest are among the top priorities to include in new parks and protected areas. Scientists fear they may not be able to document the plant species in these areas before the forests disappear.

Because the island supports a rapidly growing population, which numbered about 15 million people in 2005, there is great demand for development. Much of Madagascar's remaining rich natural environment was unprotected until recently. More than 80 percent of the island's original forest has been destroyed. In 2000, an estimated 34,000 square miles (8.8 million hectares) of Madagascar's forest remained, with about 4,479 square miles (1.16 million hectares) of the forest within the protected areas network managed by the Malagasy National Association for Protected Areas Management (known as ANGAP for *Association Nationale pour la Gestion des Aires Protégées*). The protected areas represented about 3 percent of the country's land and included eighteen national parks, five natural reserves, and twenty-three special reserves. The Malagasy Water and Forest Direction managed another 6,564 square miles (1.7 million hectares) of forest outside of the protected areas. The remaining Malagasy forest, about 23,166 square miles (6 million hectares), was unclassified by the government and considered *forêt domaniale*—national forest—vulnerable to anyone with a chainsaw or an ax.

That situation improved during the September 2003 World Parks Congress in Durban, South Africa, when President Ravalomanana proposed tripling the size of the island's protected areas. Up to 17,375 square miles (4.5 million hectares) of new protected areas, including wetland and marine environments, are included in the expansion, equal to about 10 percent of the nation's overall area. Ravalomanana reiterated his commitment to expanding his nation's protected areas during the International Scientific Conference on Biodiversity, Science, and Governance, which took place in Paris during January 2005. Making the connection between the need for sustainable development and conservation, Ravalomanana asserted that biodiversity protection is an important element in decreasing poverty and increasing opportunities for Madagascar's people. In December 2004, after extensive GIS analysis, the MBG delivered its recommendations for expanding Madagascar's environmental preserves to President Ravalomanana. It took just six months to produce the report for the Durban Vision initiative.

"This project came about because of our firm commitment and long history in Madagascar," said Trisha Consiglio, GIS analyst for the Missouri Botanical Garden. "The environmental community recognized that we were the foremost plant experts and had the best information about plant conservation."

Using decades of data

Working with the Malagasy government, the MBG used GIS to help propose new conservation areas. Base data layers for climate, topography, and geology already existed when the process began, so complex and comprehensive maps were built quickly. "We pulled in all of these years of collecting and the botanists' knowledge of the areas," Consiglio said.

Although the MBG had much of the data for the report either in-house or available from other institutions, producing scientifically sound recommendations in such a short time frame required careful and creative planning. Preparing the materials for the Durban Vision report brought together all aspects of the MBG's Madagascar research and used its connections to other international botanical institutions. Botanists, biologists, and GIS professionals in St. Louis, at England's Kew Gardens, the *Jardin des Plantes* in Paris, and in the MBG's Madagascar office, as well as research institutions in Belgium and Switzerland, all provided various pieces of background data used to create the MBG's Durban Vision report. "This project was neat because it was totally interdisciplinary," said Consiglio.

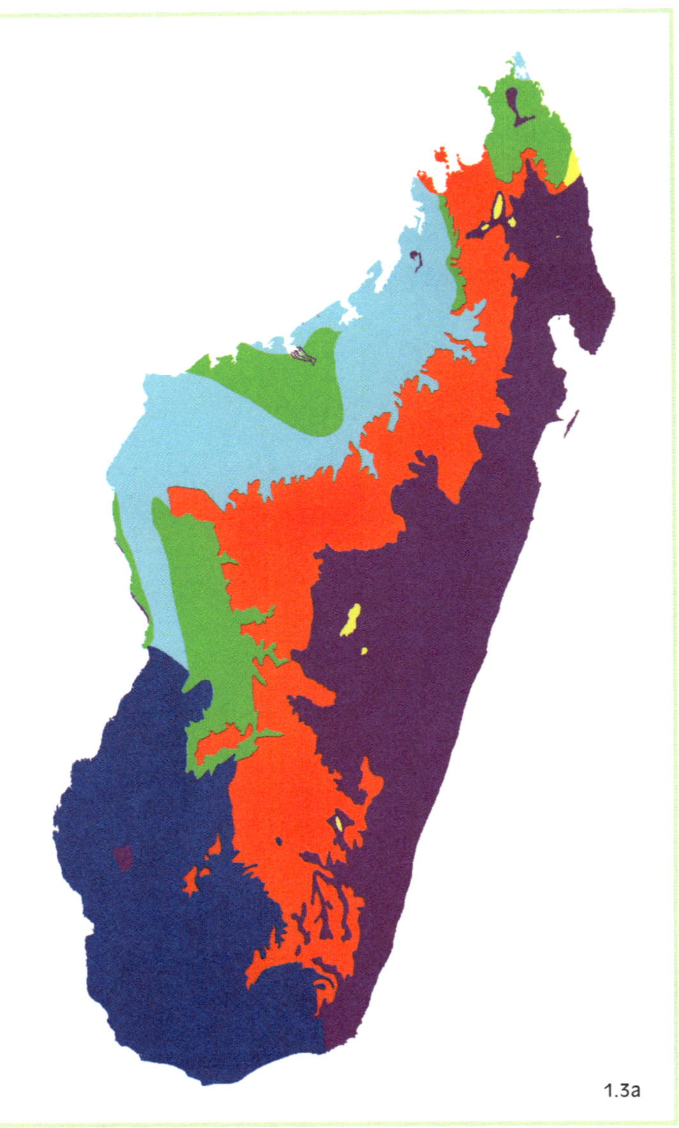

Figures 1.3a, 1.3b, and 1.3c Data layers for climate (1.3a), geology (1.3b), and elevation (1.3c) used in the modeling for the Assessment of Priority Areas for Plant Conservation project. Figure 1.3a is a refined classification of the primary vegetation of Madagascar based on the underlying geology.

Source: Missouri Botanical Garden. Digitized by George Schatz based on Cornet, A. 1974. Essai de Cartographie bioclimatique a Madagascar. ORSTOM, Paris.

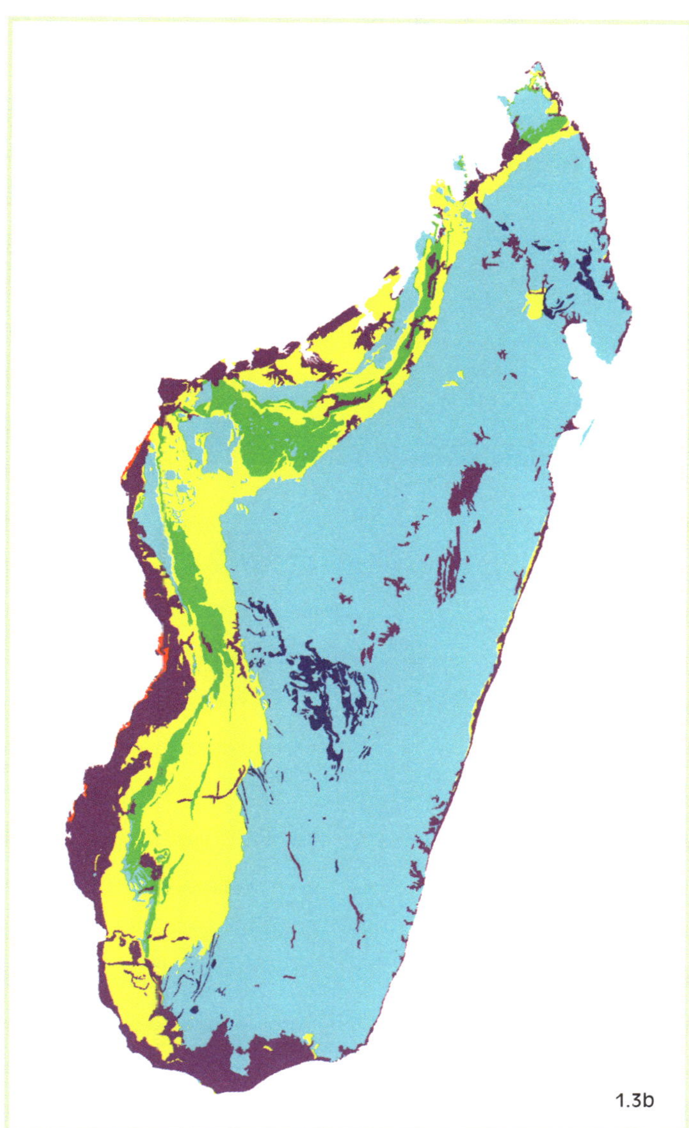

1.3b

Du Puy, D. J., and J. Moat. 1996. A refined classification of the primary vegetation of Madagascar based on the underlying geology: Using GIS to map its distribution and to assess its conservation status. In *Proceedings of the International Symposium on the Biogeography of Madagascar*, ed. W. R. Lourenço. 205–218, + 3 maps. Editions de l'ORSTOM, Paris.

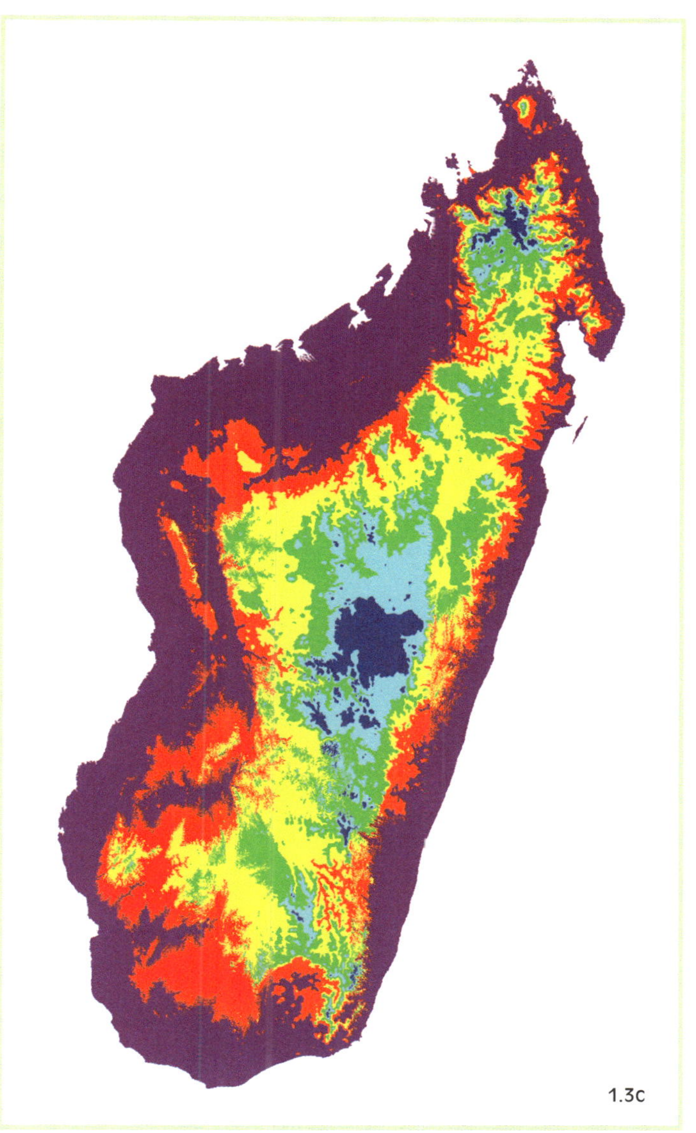

1.3c

USGS Global 30-Arc-Second Elevation Data Set. (U.S. Geological Survey, Sioux Falls, South Dakota, 1996).

Botanical brainstorming

The Durban Vision report GIS was constructed during a collaborative meeting. "What do we know and what do we need to know?" were the primary questions on everyone's mind in July 2004 when the MBG's GIS and Madagascar experts gathered in St. Louis for a brainstorming session. During two long days, botanists and GIS technicians studied paper maps and digital data layers, making decisions about what to include in the GIS and digitizing maps of areas that would be examined more closely. Combining quantifiable GIS work with the knowledge of botanists on the MBG's Durban Vision project proved to be a powerful mix.

"We came up with seventy places right off the bat that [participants] thought might be interesting, just based on the knowledge they had from working in the field and doing research," Consiglio said. "A lot of people can use the data we have and do a lot of nice things with it, but to have that expert knowledge to drive it, that's what's really important."

After the brainstorming session, Tantely Raminosoa, GIS technician for the Missouri Botanical Garden, Madagascar Program, led the two-week process of creating base data layers with the help of Paris- and Madagascar-based botanists. Raminosoa had previously conducted the GIS analysis of the Malagasy Plant Conservation Database that identified priority areas of the APAPC project.

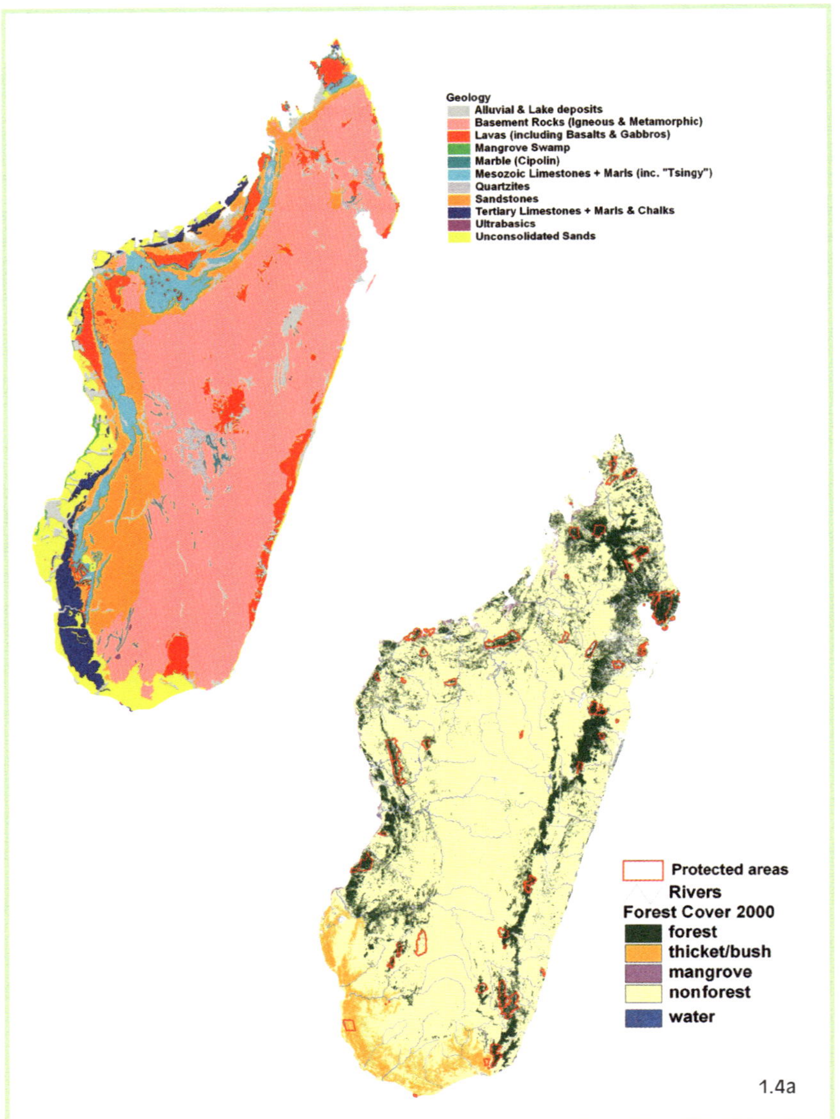

Maps by Trisha Consiglio, Missouri Botanical Garden.

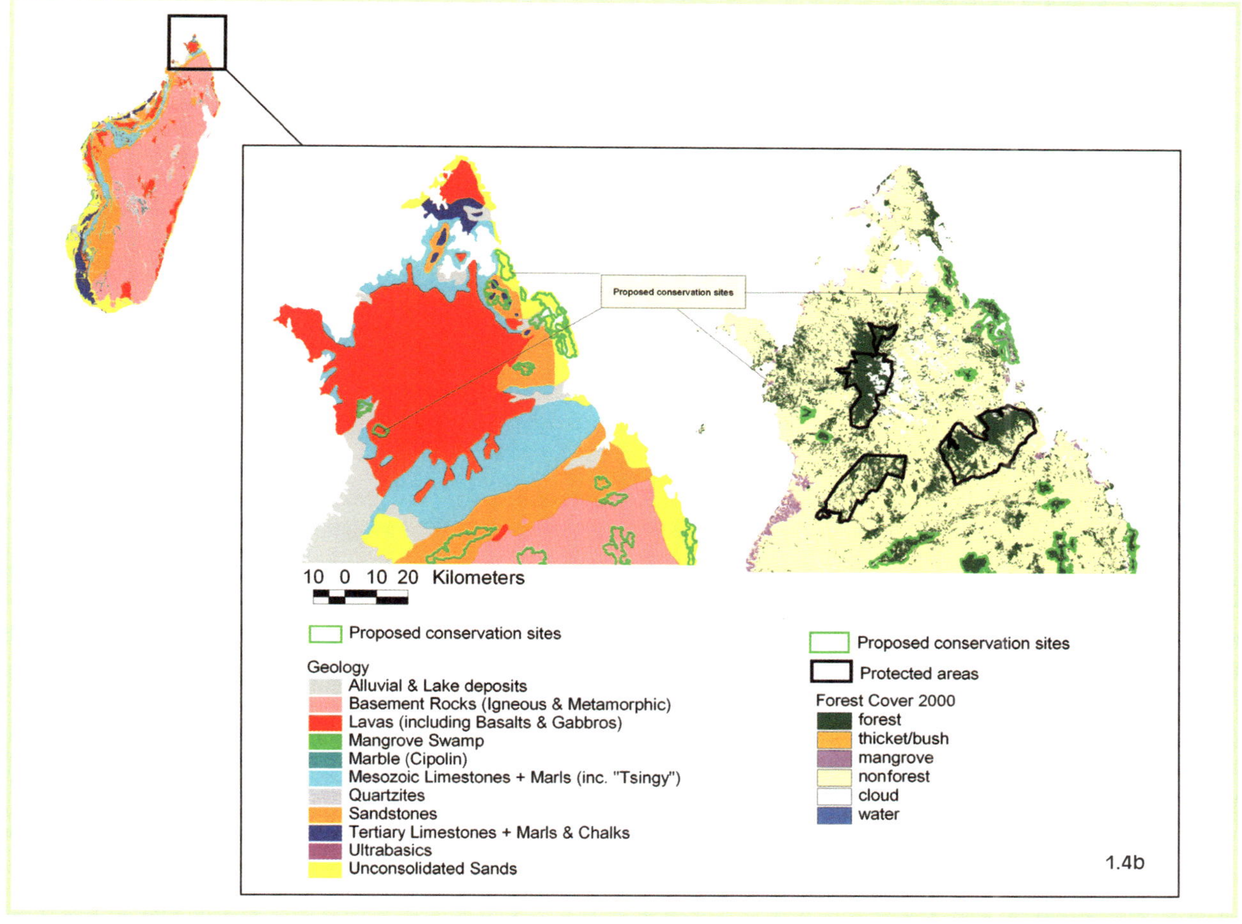

Maps by Trisha Consiglio, Missouri Botanical Garden.

Figures 1.4a and 1.4b Geology, remaining forest cover, and protected areas were used to initially determine potential sites for plant conservation. Using a gap analysis approach, proposed conservation sites were identified as those unique in geology across different bioclimatic zones, occurring on forest patches not already currently protected.

The APAPC project obtained seventy-seven priority areas for plant conservation by analyzing species and habitat. "It was a product of the combination of the intense GIS analyses and the considerable expertise of botanists. Creating the map for the Durban Vision Group is one important component of the GIS activities that are held within the APAPC project. The characteristic of its approach is the combination of the presence of significant plant species in the GIS analysis process and the expertise of botanists [in identifying] the habitat gap," said Raminosoa.

Conducting the analysis

With the data layers in place, scientists could concentrate on analysis and generating maps. The scientists used a database of approximately 1,200 species consisting of more than 15,700 georeferenced records to perform the analysis. Criteria for selecting plant species for the GIS analysis was primarily based upon endemism, the relative "representativeness" and taxonomic framework of each species. The first step was to map the primary occurrence of each species using ArcView GIS software. Once the distribution of a species was determined, project botanists verified the accuracy of the location and presence of each species.

MBG botanists conducted a species distribution analysis using WORLDMAP software. Raminosoa produced maps showing richness and range size and rarity of each species. The botanists used a spatial and tabular gap analysis to determine which species were found within and outside the current ANGAP protected areas. (Gap analysis is the scientific method of identifying and classifying components of biological diversity and determining which components are already in protected areas and which components are absent from or underrepresented in protected areas.) MBG botanists also conducted a habitat gap analysis using environmental parameters, including environmental surrogacy, climate, substrates, elevations, and vegetation type. They selected remaining areas of extant primary habitat for analysis and defined important botanic areas not included in the protected areas. Researchers used co-occurrences of remaining native vegetation, combined with a unique set of ecogeographic features, to strongly suggest where areas of local floristic richness or endemism existed, despite having little or no botanical data for those areas.

"This work is composed mostly by spatial analysis, using grid data," Raminosoa said. "They will be part of the data compilation sheet that is a full description of each conservation site."

Species distribution modeling: Predicting where plants live

MBG botanists used species distribution modeling to analyze the potential distribution of important plant species that exist outside of the current protected area network. The MBG used a modeling technique that maps similarities between areas based on ten environmental layers:

- Bioclimate
- Length of dry season
- Simplified geology
- Elevation
- Annual mean temperature
- Mean diurnal temperature range
- Temperature seasonality
- Minimum temperature of the coldest month
- Annual precipitation
- Precipitation seasonality

To improve results, the MBG integrated other prediction algorithms such as Genetic Algorithm for Rule-set Production (GARP) into its analysis. GARP is a method that automates the use of environmental data collected through field surveys to produce distribution maps and models.

Once the priority areas were defined, GIS produced basic information composed of the following seven categories for each site:

- Site identifications
- Administrative information
- Geographical information
- Botanical information
- Substrate information
- Bioclimatic information
- Vegetation information

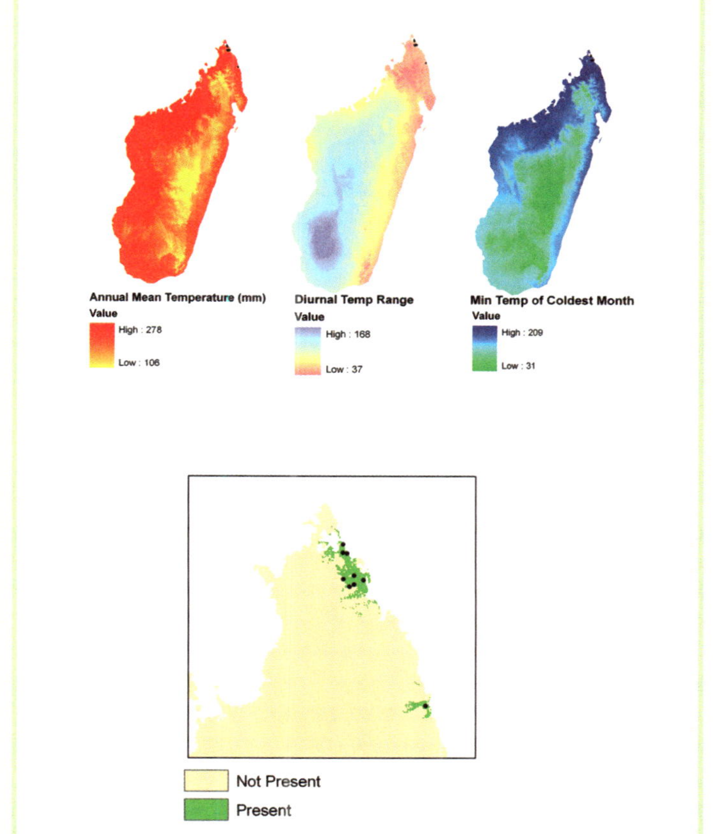

Map by Trisha Consiglio, Missouri Botanical Garden.

Figure 1.5 Climatic data layers were used to create potential distribution maps for species of conservation concern. The potential distribution of *Rhopalocarpus triplinervius* was determined by using the DOMAIN similarity algorithm applied to various climatic data layers and specimen collections of the species.

Minding the gaps

The MBG's GIS work, an integral part of its Madagascar research, extends beyond political borders and far beyond general mapping. The MBG is exploring ways that spatial modeling and remote sensing can help scientists discover where various unique plants are likely to grow. By combining GIS, prediction algorithms, and reserve selection criteria, the MBG helped devise recommendations to protect areas of high biodiversity and centers of endemism.

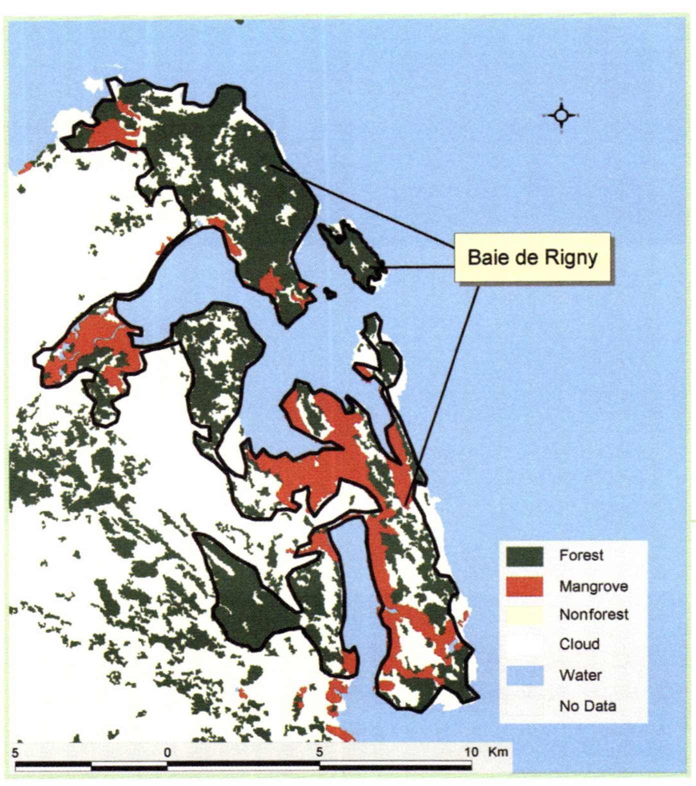

Map by Tantely Raminosoa, Missouri Botanical Garden, using data from Steininger, M. et al. 2000. *Forest Cover Fragmentation and Clearance in Madagascar.*

"We measure the distribution of plants against a number of environmental parameters such as underlying geology," Schatz said. The geological data layer exemplifies the international nature of the MBG's Madagascar GIS work. The base geology map used in the MBG analysis was produced by a team at Kew Gardens, which in turn based its work on a geology map of Madagascar produced by the French government in the 1960s. "That's been an extremely useful data layer for us to measure plant distributions against," Schatz said. Much of the work involved gap analysis of various plant species overlaid with the nation's current protected areas. Using GIS, some trends were immediately apparent.

Northern Madagascar has several large protected areas, but about forty to fifty plant species living in sandy soil and endemic to the region were not represented in the protected areas. "It was absolutely revealing to map these species against the geology and against the current protected area polygons and go 'Wow, look at that.' All of these species are occurring on sandstone, and there's no sandstone in the protected areas. It's a dramatic story that we can present to the Malagasy government and say there's this entire ecosystem that's lacking protection in the north," said Schatz.

Figure 1.6 Detail of a proposed protected area near the Baie de Rigny in northeastern Madagascar.

Malagasy researchers play key roles

The MBG's work in Madagascar has benefited the conservation of one of the world's most ecologically sensitive places. It has also created jobs and educational opportunities for Malagasy scientists and GIS professionals. Over the years, the MBG has employed Malagasy staffers to work on research and conservation efforts. Its Malagasy staff was essential in preparing the analysis and maps for the Durban Vision Group. Two Malagasy MBG Conservation and Research Program botanists, Sylvie Andriambololonera and Jeannie Raharimampionona, are coordinators of the APAPC project. Graduates of the University of Antananarivo's botany department, Andriambololonera and Raharimampionona have worked for the MBG Madagascar program since 1992. They also participated in weekly working sessions of the Durban Vision Group, focusing on the technical and scientific aspects of the process. The MBG was the primary botanical institution within the Durban Vision Group and led its flora subgroup.

"From the start of the process, MBG viewed the Durban Vision as a great opportunity to integrate the APAPC project's results in a conservation zone as a whole," Andriambololonera said. "Therefore, since the beginning, MBG has been fully involved in the implementation of the Durban Vision process." Raminosoa brought together the Durban Vision Group's GIS experts to produce and refine the preliminary maps of potential new protected areas. This work builds on the World Wildlife Fund's analysis of conservation priorities for Madagascar's ecoregions.

Photo by David Rabehevitra, Missouri Botanical Garden.

Figure 1.7 Preservation of unique forest habitats, such as this one in the Antaimby area of Madagascar, is one of the highest priorities for the Malagasy government and conservation organizations.

Future research and analysis

With the Durban Vision Group's mapping process in its final stages, work on the APAPC project continues. "We are now conducting the modeling of species distribution (those falling outside the current protected areas) using different algorithms," Raminosoa said.

Collaborating with different groups on the Durban Vision process created a demand for the MBG's GIS services. "As MBG is becoming more involved with consultations with other organizations, we are furnishing our services based on our competence, including GIS analyses activities of the botanical collections database," he said.

Raminosoa also leads the GIS activities of the MBG's consultancy with the Ambatovy Mining Project and works with researchers, specialists, and students on their GIS analysis projects. He continues to focus on the Priority Areas for Plant Conservation project as well as other GIS analyses related to different projects of the Missouri Botanical Garden, Madagascar Program. Thanks to GIS, the MBG was able to present a very real and compelling scientific report that will help realize the Durban Vision of expanding the nation's protected areas and preserving one of the world's most biologically diverse places.

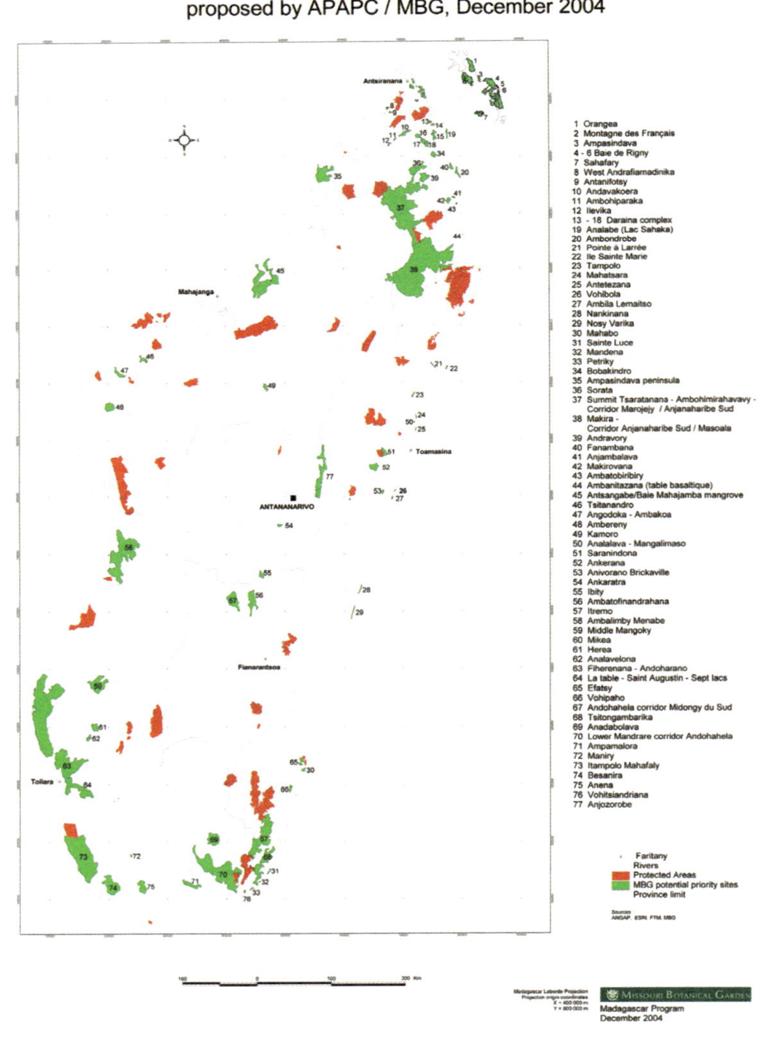

Map by Tantely Raminosoa, Missouri Botanical Garden.

Figure 1.8 The final map of proposed new biologically protected areas in Madagascar, December 2004. It was developed by creatively using GIS and twenty-five years of the Missouri Botanical Garden's research on the island's flora.

References

African Conservation Association. Madagascar. www.africanconservation.com/madagascar.html

Conservation International. Biodiversity Hotspots. www.biodiversityhotspots.org/xp/Hotspots/hotspotsScience

Madagascar Wildlife Conservation. www.mwc-info.net/en/index.htm

Missouri Botanical Garden. A View from the Top, Using Aerial Photography to Measure Madagascar's Littoral Forests. www.mobot.org/MOBOT/research/littoral/welcome.shtml

Missouri Botanical Garden. Projects in Africa, Madagascar. www.mobot.org/MOBOT/Research/africaprojects.shtml

Preston-Mafham, Ken. 1991. *Madagascar: A natural history*. New York: Oxford.

Royal Botanic Gardens, Kew. GIS and Conservation for Madagascar. www.rbgkew.org.uk/gis/projects/madagascar

State University of New York, Stony Brook, Institute for the Conservation of Tropical Environments. icte.bio.sunysb.edu

Pollution 2

Settling the dust along the international border

The international border towns of Douglas, Arizona, and Agua Prieta, Mexico, shared a serious dust problem that threatened public health.

Data provided by 2004 ESRI Data & Maps.

Dust kicked up by hot, dry winds cutting across the open desert along the international border between Agua Prieta, Sonora, Mexico, and Douglas, Arizona, began threatening public health in the 1990s. Potentially harmful dust levels exceeded state and federal standards in Douglas. A bilateral solution to the problem was important since Agua Prieta was the fastest growing city on the U.S.–Mexico border. Arizona public health officials first needed to pinpoint where the dust came from and then determine how to clear the air. The Arizona Department of Environmental Quality (ADEQ) sought an accurate way to find, collect, and analyze geospatial data on air quality and pollution sources.

Source: T. S. Summers, GeoEye.

In 1998, Tom Summers, GIS project manager for the ADEQ, developed geospatial data for the state's emission inventories—lists of communities' air pollutant sources and amounts. Such inventories help policy makers establish emission standards. "Emission inventories look at land use and activities that might cause some harm to the environment," Summers explained. "For instance, dirt roads with bikers running across them stirring up lots of dust into the atmosphere." Analyzing emission inventories with GIS allowed air quality officials to see exactly where the various sources of dust were concentrated. Overall, the GIS project for Douglas–Agua Prieta was a success and became a template for other projects.

When Summers started working on the emission inventories, he had only low-resolution aerial imagery and a second-hand Sun Microsystems workstation available to him, yet he vastly improved how his agency gathered and analyzed data. At that time, ADEQ workers in the field wrote notes on paper maps and then used a spreadsheet to compile the information for state officials.

Figure 2.1 This 1-meter-resolution multispectral IKONOS satellite photo shows Douglas, Arizona, and Agua Prieta, Mexico, in the winter of 2000. This was used as the base image for the emissions inventory research and covers a 33-square-mile area straddling both sides of the U.S.–Mexico border. Douglas is at the top of the photo, and the grid-like street layout of Agua Prieta is at the bottom.

Imagery worth thousands of words

Once computers improved and affordable high-resolution satellite imagery became available, Summers could conduct the kinds of GIS analysis he had been unable to accomplish with the older technologies. ARC/INFO Workstation,

Figure 2.2 Separate bands of the same image—red, blue, and green—were stacked on a 4-meter-resolution multispectral image to create a full-color 1-meter-resolution image.

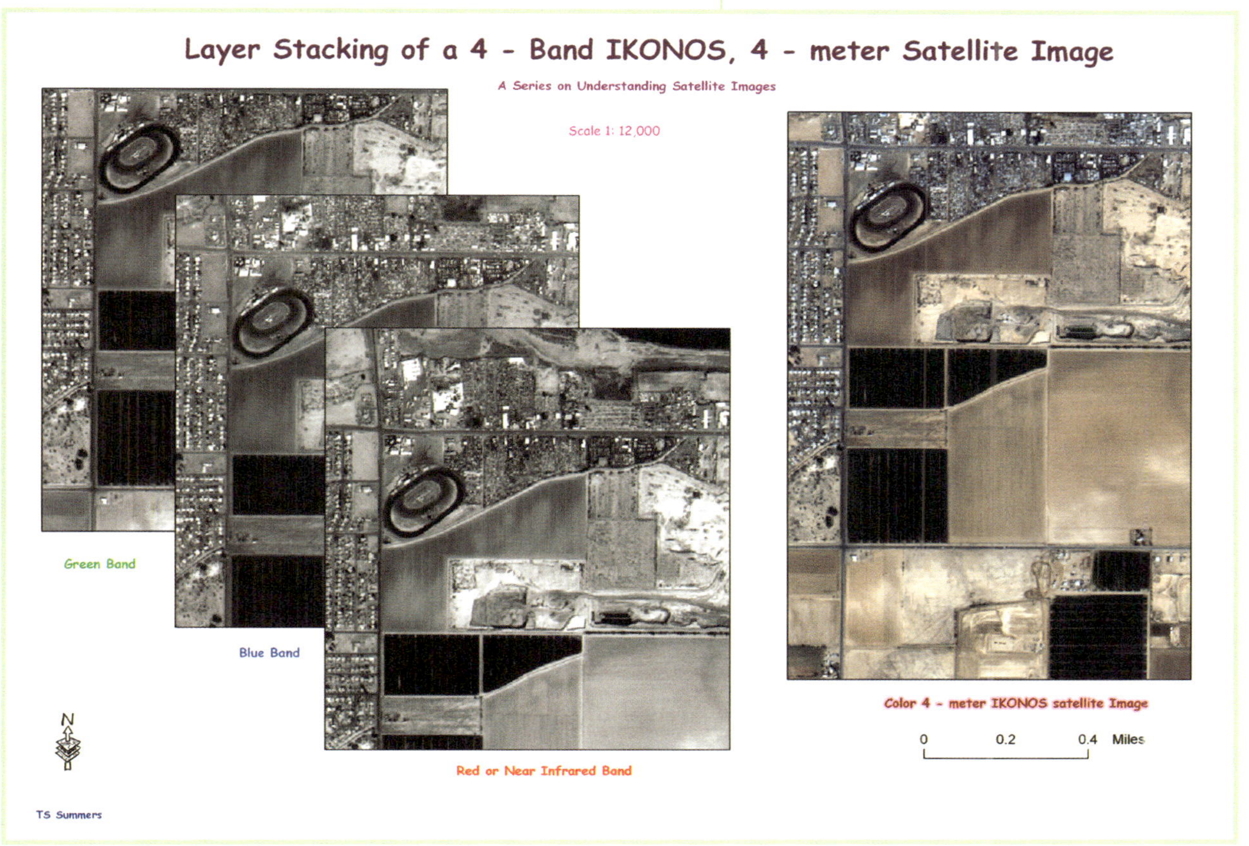

Source: T. S. Summers; GeoEye.

ArcEditor, and ArcGIS Geostatistical Analyst software, high-resolution satellite photos, and later, Tablet PCs running ArcEditor software allowed the agency's staff to digitize and analyze data in the field. Summers dramatically enhanced his agency's field efforts and reduced the time to create the emission inventories. "If you have your geospatial data and can combine it with imagery, that's worth thousands of words," Summers said.

The ability of GIS to combine geospatial data with high-resolution imagery provided agency officials with improved decision-making tools. Summers found imagery to be a great tool to view how people deal with the environment. However, when Summers began working with GIS in the early 1990s, his imagery options were limited to 30-meter Landsat or 25-meter SPOT multispectral and 10-meter panchromatic black-and-white photos. He found these tools worked fine for larger, regional environmental studies, but didn't suit projects in urban areas. Aerial photos were a better resource, but were expensive. During the mid-1990s, the ADEQ manually studied the effect of land use on air quality by putting a piece of clear plastic with a grid over an aerial photo to determine what activities were happening in each grid square.

Maps for management

In 1998, the ADEQ's air-quality group recruited Summers to develop GIS projects. Management's support for developing GIS databases was lukewarm at first, but the power of maps and imagery helped convince them of the value of GIS. "Managers found that the databases and the maps I produced greatly increased the accuracy and understanding of their regional air-quality studies," Summers said. "Over time, the need grew even higher for these products to a point management could not do without them."

Summers used ARC/INFO and an aerial image to construct a test GIS land-use database for developing an air-quality emission inventory. At that time, commercial high-resolution satellite images weren't available, so Summers used an old color aerial photo of an area in Phoenix to develop a case study to sell the idea to his bosses. "I explained to management that in January 2000, Space Imaging was going to release new 1-meter, high-resolution satellite imagery (IKONOS) and we should get it for the air-quality emission inventory," Summers said.

Figure 2.3 Once the 1-meter image is ready, it is overlaid with a 500-meter grid and indexed for easy reference during field survey work. The index grid is also used in the emissions modeling process to predict or show where and how pollutants are being released into the air. This index grid image is of the Salt River area of Phoenix, another area that underwent the emission inventory process to help determine sources of air pollution.

Salt River Index Map

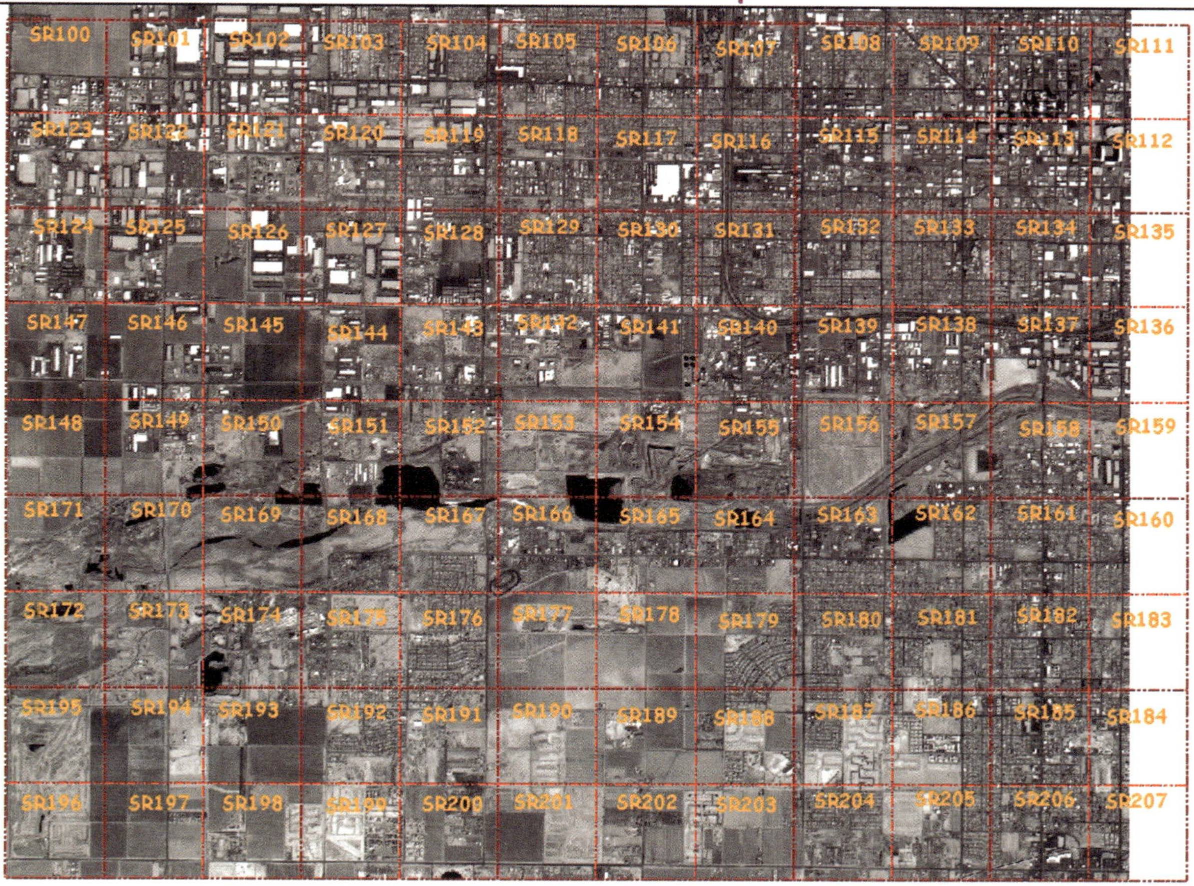

Source: T. S. Summers; GeoEye.

Getting the big picture

Management was reluctant at first, but agreed to give Summers $2,000 to purchase the first high-resolution satellite image. Working with Global Systems Modeling Ltd. of Tucson, Arizona, the ADEQ purchased 1-meter panchromatic and 4-meter multispectral data for Summers's 33-square-mile pilot study of the area in and around Douglas and Agua Prieta.

The ADEQ wanted to use GIS to analyze detailed information about the international border's serious dust problem in an effort to literally clear the air. Once Summers had his satellite photo, his first task was dividing the area into a grid to produce maps big enough to take notes on, but small enough for easy handling in the field. The field maps' resolution needed to show as much detail as possible without becoming pixilated. The printed field maps measured 17 by 21 inches with each tile covering about 1,000 meters on a side.

One drawback of satellite images of Arizona, Summers found, is that extremely bright sunlight reflecting from the many light-colored roofs can make it difficult to pick out some details. He and his team experimented with the color histogram and discovered that 11-bit data with 2,048 shades of gray increased the color depth of the image better than 8-bit data with 256 shades of gray. The 11-bit data helped reveal details that would have been lost in a color-reduced version of the image.

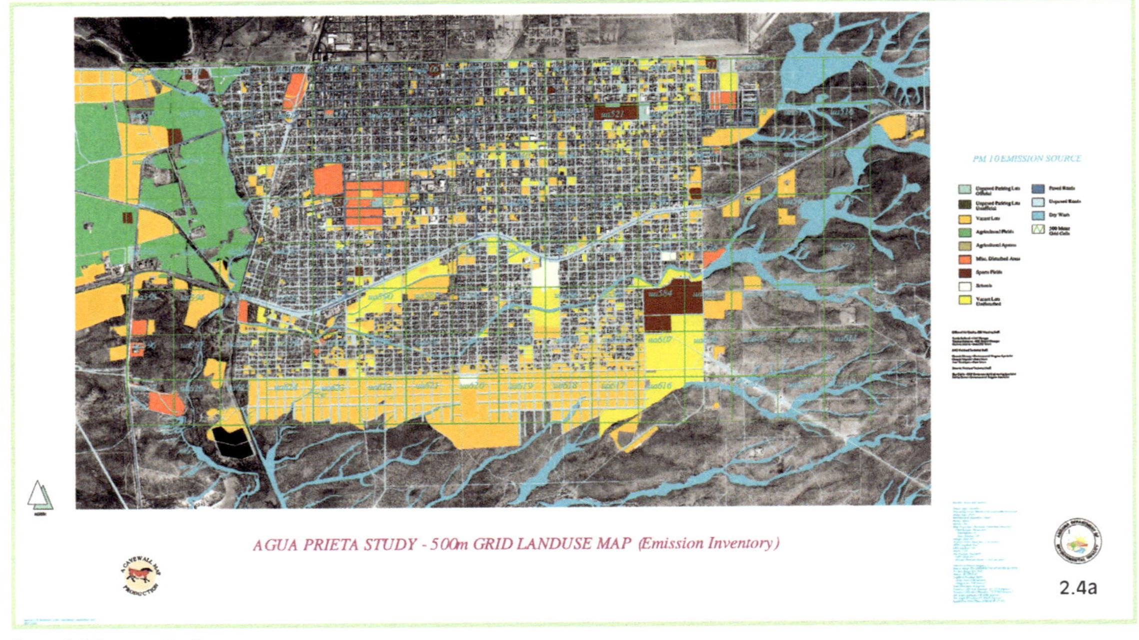

Source: T. S. Summers; GeoEye.

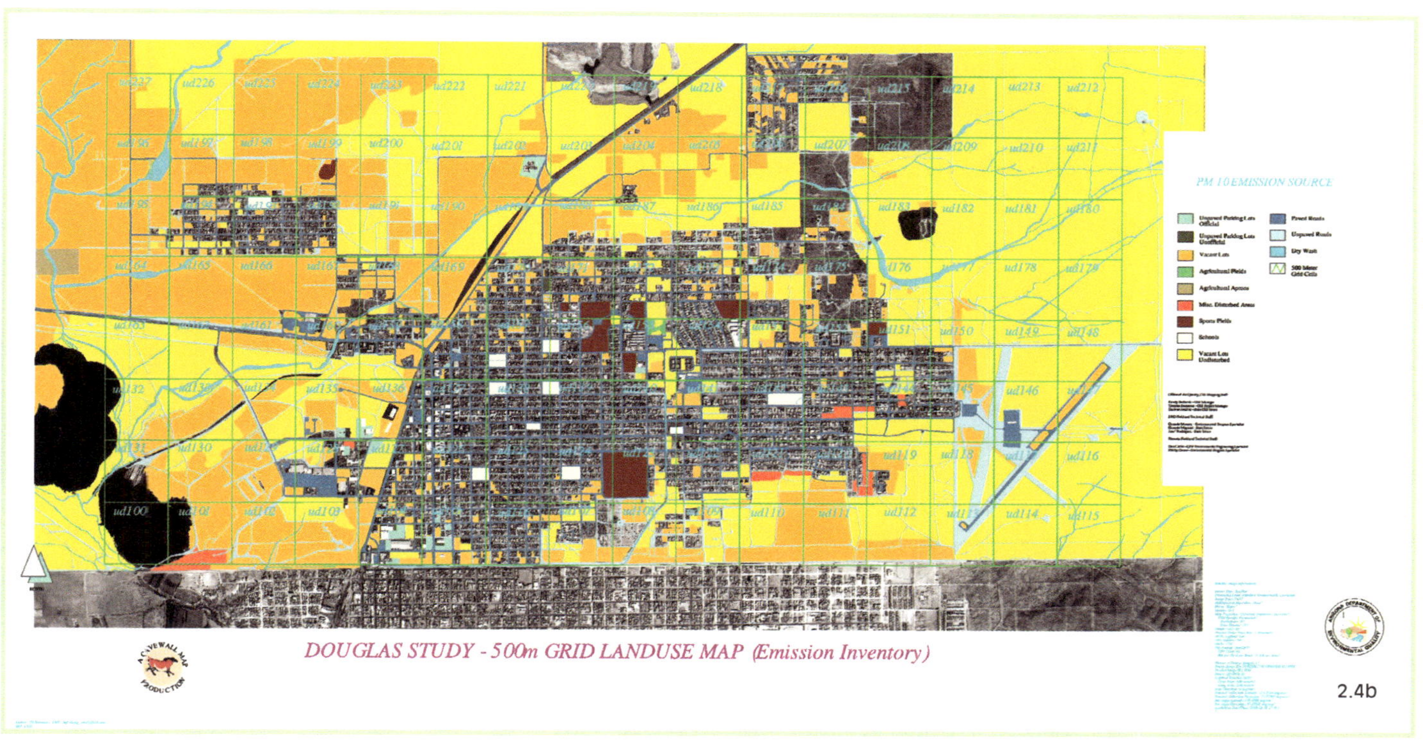

Source: T. S. Summers; GeoEye.

Figures 2.4a and 2.4b Layers showing the various types of land uses associated with the emission inventory as well as drainage and streams. Images overlaid with the indexed 500-meter grid are added to both Douglas and Agua Prieta. The grid allowed researchers to determine how much of each type of potential emission source existed in a given area.

An index map and a model method

An overall index map showed the location of each map tile within the study area. Developing the index not only allowed for creating the field maps, but also maps for use within a model. Modeling, perfectly suited to emission inventory work, takes information from a grid, such as data gathered by field workers, and relates it to various types of activities.

The ArcEdit function was used to digitize streets, the edges of pavement, agricultural fields, and other features directly over the satellite image. It was important to Summers for the study to identify streets with curbs, unpaved streets, vacant lots, paved and unpaved parking lots, and other dust generators.

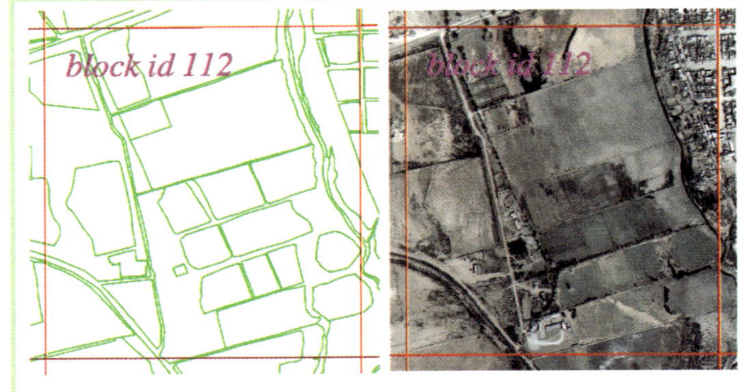

Source: T. S. Summers; GeoEye.

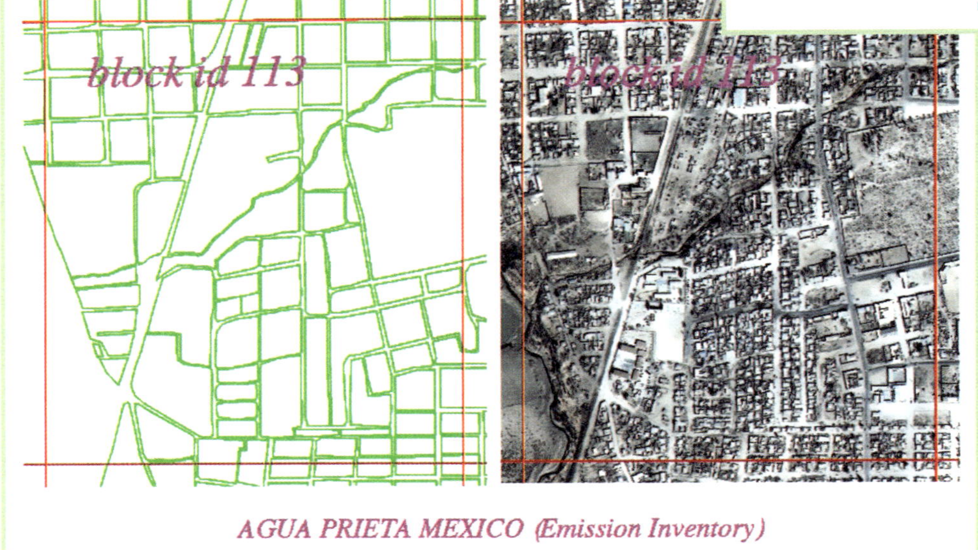

Source: T. S. Summers; GeoEye.

Figures 2.5a and 2.5b The larger 500-meter grid image is broken down into individual blocks and the detail stripped down to better show the borders between features on the ground. These individual grid blocks can then either be printed out or loaded onto Tablet PCs and given to a field researcher who then observes and verifies land-use activities on the ground.

Hitting the dusty trail to clear the air

Field agents then fanned out through the area in pursuit of hazardous particulates in the air. They gathered georeferenced data on emissions sources, such as paint factories, restaurants, auto body shops, dry cleaners, electronics, and other manufacturing facilities. Researchers also gathered data on home heating and cooking fuels, wood burning, and agriculture.

GIS revealed that Agua Prieta's unpaved streets were a major source of airborne dust. In February 2003, the North American Development Bank funded a loan to pave 21 miles of roads in Agua Prieta.

Air-quality studies along the international border and other studies in the Salt River area of Phoenix continued through 2005, producing numerous positive results. The state, county, and city planned to develop new regulations to control dust from sand and gravel mining operations in the Salt River area. The governments also planned installation of greenbelts in the riverbed to reduce airborne pollutants.

Tablet PCs tame field data

When Tablet PCs appeared on the market, Summers quickly realized the potential for improving the productivity of gathering data in the field. "I understood in 2000 that there was a need for a tablet-type PC for field work, and when they appeared on the market I realized this is what we were looking for," Summers recalled.

Field maps could now be loaded onto Tablet PCs. Coverages were first created in ArcGIS, and a basemap was set up using the satellite imagery and the study area boundaries. Adding a grid to the basemap was the final step, and then the map could be loaded onto a Tablet PC. Using the Tablet PC loaded with a digitized map, a worker in the field using ArcGIS could assign attributes associated with various types of land uses. The Environmental Protection Agency (EPA) developed a system that assigns numbers to different types of land uses. Coverage areas were overlaid and combined with the grid. A worker in the field using ArcGIS Desktop software could conduct preliminary on-site analysis by setting up the grid to display the percentage of land within a grid associated with a certain land use.

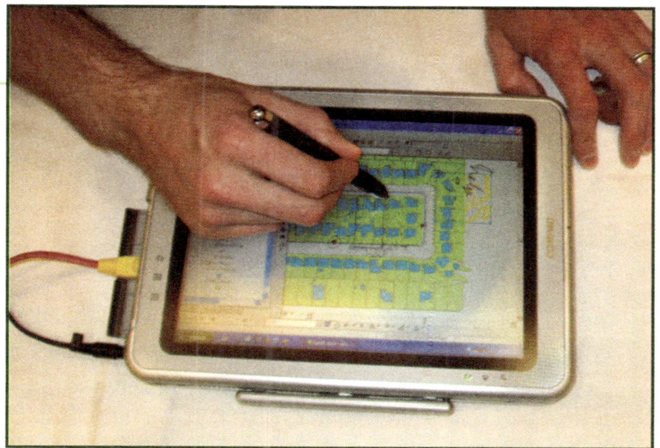

Source: T. S. Summers; GeoEye.

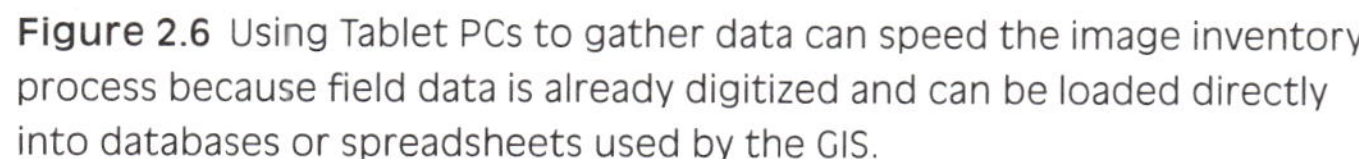

Figure 2.6 Using Tablet PCs to gather data can speed the image inventory process because field data is already digitized and can be loaded directly into databases or spreadsheets used by the GIS.

"The geodatabase and table computer is the next step in the evolutionary scale," Summers said. "Now you can actually have people out in the field using imagery as the background digitizing these land uses right there in the field, making their notes, and the whole thing can be downloaded to the geodatabase."

Tablet PCs have eliminated what was once one of the slowest parts of the emission inventory process—transferring data and field notes—while reducing the potential for introducing errors. Before the Tablet PC emerged, an emissions inventory project might have two hundred or more individual paper maps for use in the field. Transferring information written on the maps into the geodatabase was done by hand.

Now, field survey data is sorted by the GIS and georeferenced using the indexed satellite image. ArcGIS Desktop's database table creates the polygons representing the various EPA land-use codes. After uploading the field data to an office computer, the field surveyor edits the shapefile and polygon information, places the grid back on top of the basemap, and adds the new information into the model. Finally, the surveyor merges the EPA models for analyzing emission factors and land uses with the ADEQ's model.

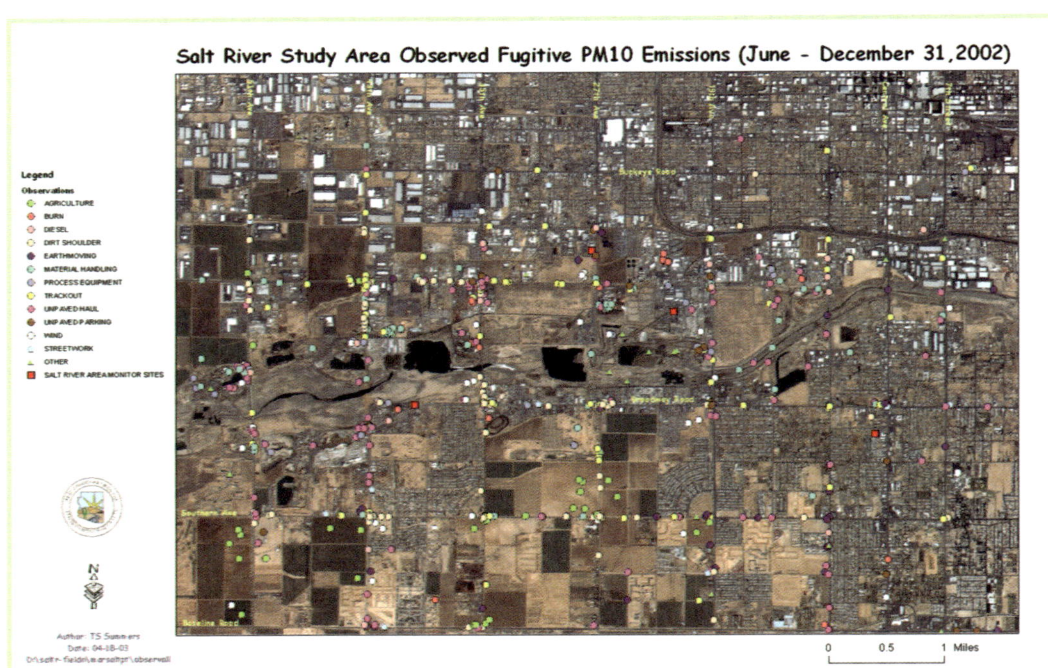

Source: T. S. Summers; GeoEye.

Figure 2.7 Field survey data is sorted by the GIS and georeferenced using the indexed satellite image. The result shows the exact location of activities generating high levels of dust. These images show an emission source inventory for the Salt River area of Phoenix.

The right tools and quality data

Summers has learned from experience that spending more on technology and data acquisition at the beginning of a project will create a better result in the long run. "It is important to understand that the quality of what you put in will affect the quality of what you get out."

Although GIS technology has become more sophisticated during the past several years, knowing how to best apply the tools available for the job at hand remains the key to creating a successful GIS project. Summers observed, "As the software expands, your abilities are much greater, but you need to be aware of all of the different tools out there and how they can be used together in the process. As the tools get better, we're only going to be able to do more with them and right now you're only limited by your imagination."

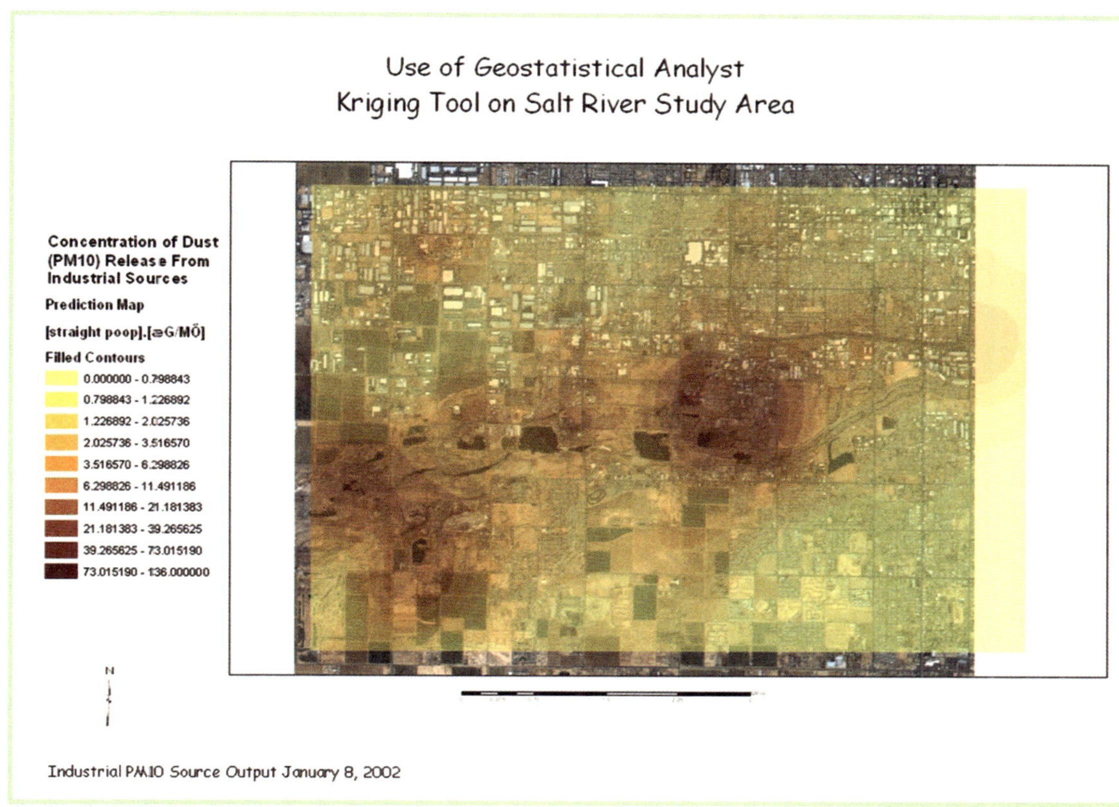

Source: T. S. Summers; GeoEye.

Figure 2.8 This map shows the result of modeling of field data based on the grid system and interpreted through a process known as kriging, an interpolation technique in which measured values are weighted to derive a prediction for an unmeasured location. In this case, the map shows the predicted emission of dust from industrial sources such as construction and open-pit sand and gravel mining in the Salt River area of Phoenix.

GIS tracks contamination from the World Trade Center collapse

In the wake of the terrorist attacks of September 11, 2001, the Environmental Science Center at Syracuse Research Corporation (SRC) analyzed data on contaminants in indoor air and dust in lower Manhattan. SRC, working for Region 2 of the Environmental Protection Agency (EPA), used quantitative geographical analysis methods to determine if any spatial pattern existed in the locations where contaminants exceeded health standards. SRC tested the hypothesis that these locations did not exhibit a geographic pattern relative to the World Trade Center site, which might indicate the cleanup of the residences was not effective.

Source: Syracuse Research Corporation.

Figure 2.9 The study divided the area of lower Manhattan south of Canal Street near the World Trade Center site into forty-five statistical summary areas (SSAs), created by combining census block groups. The cross-hatching in this map indicates areas where data was not collected.

"The EPA came to us and wanted to see if there was any real geographic pattern of asbestos contamination following the cleanup work," said Dr. William Thayer, a professional engineer with the Environmental Science Center in Syracuse, New York. The spatial analysis focused on the concentration of indoor airborne asbestos in residential buildings. The study area, or spatial scale of the study, included lower Manhattan below Canal Street. The spatial analysis was performed at three levels of spatial resolution:

- Building level (highest level of resolution available)
- Statistical summary area (SSA) level
- Site level (lower Manhattan)

SRC used two types of point-pattern analysis, nearest-neighbor and Ripley K analysis, and a spatially filtered Poisson model for the study, which used ArcView and MapObjects software. Point-pattern analysis is a way of studying the location of events and revealing facts about the distribution of those locations. Point-pattern analysis can show whether the locations of events are randomly clustered or regularly distributed. Nearest-neighbor analysis examines the distances between each point and the closest point to it. The second type of point-pattern analysis used in the study, Ripley K analysis, preserves distances at multiple scales and can quantify the intensity of patterns at multiple scales.

According to Thayer, it is important to analyze the data at different resolutions. For example, decreasing resolution by aggregating or averaging the data over larger areas results in some loss of information. However, aggregating may reveal geographic patterns hidden by data fluctuations in data over small distances. Three characteristics of the asbestos data made the spatial statistical methods SRC employed more effective than the standard nonspatial statistical methods such as t-tests, which are statistical methods used to compare two groups:

- In 96 percent of the cases, the asbestos concentrations were so low an exact number could not be assigned to the sample.
- Less than 1 percent of the samples exceeded the health-based standard for asbestos.
- Samples were only collected after cleaning the building, so there was nothing to compare the results to.

In this case, GIS analysis found no spatial pattern of excessive asbestos levels in the area around the World Trade Center during the post-September 11, 2001, clean-up efforts.

The study divided lower Manhattan into forty-five statistical summary areas (SSAs) created by combining census block groups. Residences that were cleaned and tested were

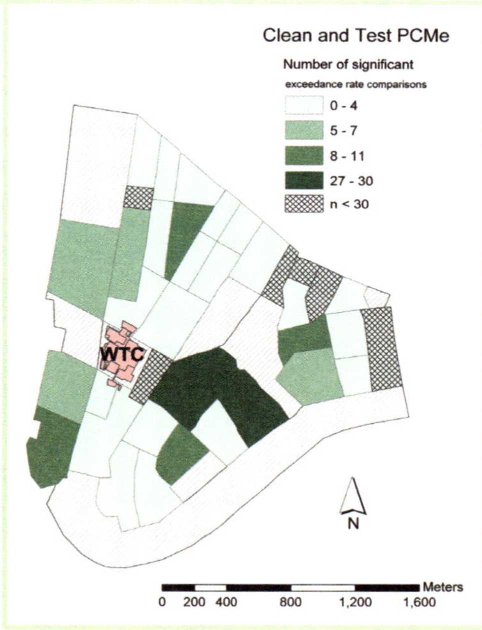

Source: Syracuse Research Corporation.

Figure 2.10a This map is based on data from residences that were cleaned and tested for contamination after September 11. This analysis revealed three statistical summary areas east of the World Trade Center with excessive airborne asbestos rates significantly greater than the other study areas.

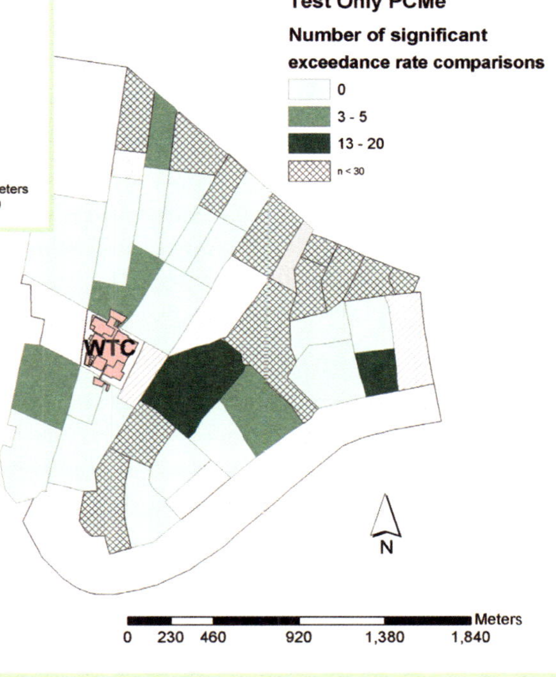

Source: Syracuse Research Corporation.

Figure 2.10b This map is based on data from residences that were not cleaned but tested for asbestos contamination. The data shows two statistical summary areas east of the World Trade Center and other SSAs in the northeast corner of the study area with excessive airborne asbestos rates greater than the other study areas.

located in thirty-six of the SSAs, and residences that were tested, but not part of the clean-up program, were located in thirty-eight of the SSAs.

The air sample analysis did find some locations where asbestos was at elevated levels, but the GIS analysis revealed no obvious spatial pattern in the locations. The GIS analysis also determined that higher concentrations of asbestos could be explained by the location of the sampled buildings. Areas with higher sample numbers contained proportionately more locations with high asbestos concentrations, as expected. Locations with high amounts of airborne asbestos varied across all of lower Manhattan, and no obvious pattern relative to the World Trade Center site existed.

Other uses of point-pattern analysis can help uncover connections between certain types of pollution and harmful effects upon people and the environment. The EPA has collected large amounts of data through its Toxic Release Inventory (TRI) program. Thayer is interested in using point-pattern analysis of TRI data to detect relationships between possible health effects or environmental impacts versus various types of emissions.

Thayer wants to explore the possibility of creating an automated data mining process that could analyze toxic release inventory data using point-pattern analysis. For example, the automated system could show possible correlation between cancer rates and high releases of certain chemicals. This type of analysis could then allow any follow-up studies to be more focused.

Dan Griffith (University of Texas), Gary Diamond (Syracuse Research Corporation), and Brad Allen (EcoScience Corporation) also contributed to the spatial analysis of the WTC data.

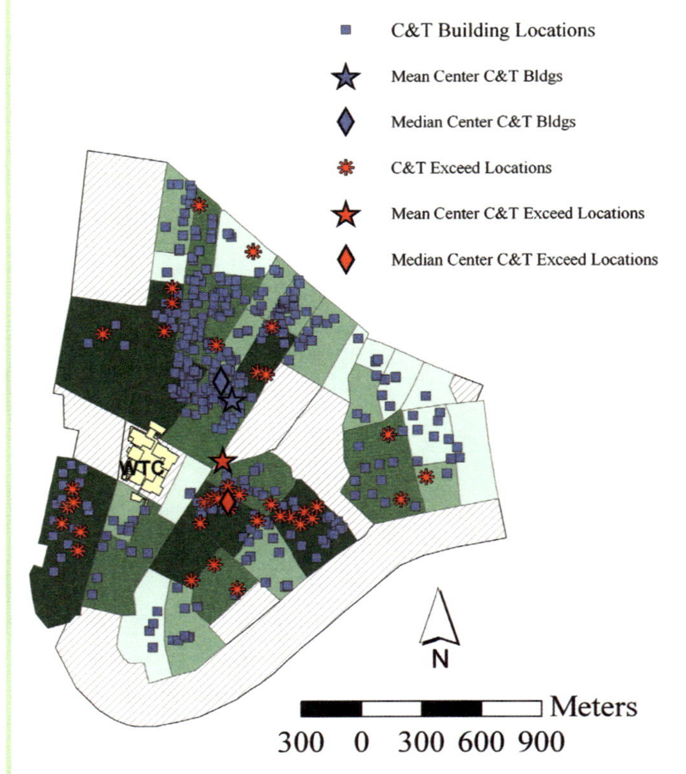

Source: Syracuse Research Corporation.

Figure 2.11 This map indicates no spatial pattern between the World Trade Center site and the buildings where the health-based standard for airborne asbestos was exceeded, once the samples collected from each building were analyzed. Statistical summary areas are shaded to indicate the number of samples collected, with the darkest shade of green indicating areas with the most samples. A strong relationship exists between the sample size and the locations where the health standard for airborne asbestos was exceeded.

References

Diamond, Gary, Daniel Griffith, and Bill Thayer. 2004. *Geospatial analysis of WTC dust clean up program data.* Syracuse, N.Y.: Syracuse Research Corp. PowerPoint presentation.

Kolbe, Jim (Congressman). February 19, 2003. Douglas to finally receive relief from air pollution, announces Kolbe. Press release.

Shields, Barbara 2005. Dust mapping using GIS and satellite imagery. *Environmental Observer.* (Winter): 6–7. www.esri.com/library/newsletters/environment/envobs-winter2005.pdf

U.S. Environmental Protection Agency. March 29, 2005. EPA response to September 11, 2001. www.epa.gov/wtc/index.html#testdata

Coastal zone management 3

Conserving California's underwater forests

Valuable kelp forests run along the entire coast of California.

Data provided by 2004 ESRI Data & Maps.

Most people dismiss it as seaweed, yet kelp is one of California's most important coastal resources. California's coastal kelp "forests" are vital habitats for sea life as well as a cash crop. These kelp forests, or beds, play an important role in the life cycle of several fish species, including rockfish and lobster. In some coastal areas, commercial harvesters collect kelp to produce additives in salad dressing, toothpaste, desserts, lotions, cleaners, and paint. Abalone farmers also lease a few of the state's kelp beds.

The state divides the kelp beds into eighty-eight administrative areas for conservation management purposes and to control leasing to the commercial harvesters and abalone farmers. The California Department of Fish and Game (DFG) supervises these administrative kelp beds. To manage this vast and valuable resource, the DFG uses GIS to maintain a digital kelp information archive. The data sources and techniques include aerial photography, satellite imagery, scanning, and georeferencing as well as custom GIS programming and documentation. "We took the Department of Fish and Game's kelp mapping into the digital age," said Paul Veisze, DFG's regional GIS coordinator and project manager for kelp-bed mapping.

Figure 3.1 Kelp frequently breaks free from rock moorings offshore and washes up on California beaches. It's often considered "seaweed," and its true value as a wildlife habitat is overlooked.

Surveying underwater forests

The DFG's primary goal for a 1999 kelp study was to create baseline data for assessing the effects of current and future uses of coastal kelp. Other objectives included digitally remeasuring the printed kelp maps created in 1989, establishing efficient digital methods for assessing kelp beds, and creating a digital archive of kelp information. The nature of California's coastline and its kelp beds made for a challenging GIS project. Kelp beds consist of long strands of brownish green plants anchored to the sea floor. Accounting for clouds, tides, waves, and wind were important while taking and analyzing the aerial photos. It was also important to capture the extent of the kelp stands.

Kelp usually grows in a relatively narrow zone within a few hundred yards of the coastline. One of California's largest kelp beds extends more than 2 miles offshore from the northwest corner of San Nicolas Island. Along the mainland coast off Point Loma, just outside San Diego Bay, the huge kelp bed extends more than 1 mile offshore. Most other beds are much smaller, tucked in covers all along the shoreline.

Kelp beds are a rich habitat for numerous invertebrate and fish species. The beds are vital in the development of young rockfish while also providing prime feeding grounds for sea otters and several types of birds.

Photo by Robert Scally.

Figure 3.2 The 77,000-gallon Birch Aquarium kelp tank at the Scripps Institute of Oceanography in La Jolla, California, allows us to see what kelp forests and the abundant marine life they support look like underwater.

Just as terrestrial forests may vary in tree composition, studies have found that kelp beds also show regional variations, according to Veisze. Giant kelp, *Macrocystis pyrifera,* populates Southern California's kelp beds while bull kelp, *Nereocystis luetkeana,* dominates northern California's beds. Kelp beds along California's central coast are a mixture of both types. Until recently, kelp bed area figures derived from printed maps with hand-drawn features determined harvesting regulations. In 1999, the DFG began to revise a 1996 environmental document supporting kelp regulations. Upgrading the DFG's kelp bed maps was a key part of the revision.

Creating new maps from old

California's kelp surveys took place over several decades, the first in 1912. But before 1999, the only statewide kelp forest mapping was the 1989 California Coastal Kelp Resources Survey, conducted by the DFG and state contractor ECOSCAN Resources, Inc., of Watsonville, California. This survey produced a set of paper maps with the kelp bed boundaries delineated by hand tracings from aerial photos. During the 1989 mapping project, survey crews used color infrared slide film and a 35-mm Nikon camera for the aerial photos.

Figure 3.3 Marine mammals, such as these endangered sea otters, depend on kelp beds as a source of food.

With the photos as a base, outlining and inking the kelp canopies onto overlays involved projecting the color infrared slides onto 1:24,000 USGS topographic maps. The next step was to scan, scale, and compile the hand-drawn information on the overlays. The resulting set of eighty-four paper maps included the topographic base and showed shorelines, watercourses, rocky tidal zones, offshore rocks, islands, sand beaches, some selected place names, and the DFG's official administrative kelp bed numbers. "It was a manually intensive process, but it was reasonably accurate," Veisze said.

The 1999 kelp survey began with the 1989 maps as a base. DFG scanned the original 24-by-36-inch maps into packbit compressed TIFF-format images and georeferenced to USGS topographic basemaps using a customized ARC/INFO application developed by ESRI. Clipping kelp canopy features from the map images allowed for conversion into the ARC/INFO ArcGrid format. ESRI assisted the DFG team members by providing a digital workflow that enabled them to scan the 1989 maps and put them into a real-world coordinate system. "We scanned [the] maps and then digitally stripped away everything that was not kelp," Veisze said.

Figure 3.4 A kelp bed surface canopy map, converted from scans of paper maps into the ARC/INFO ArcGrid format, then depicted as an image overlay to administrative kelp beds and county boundary.

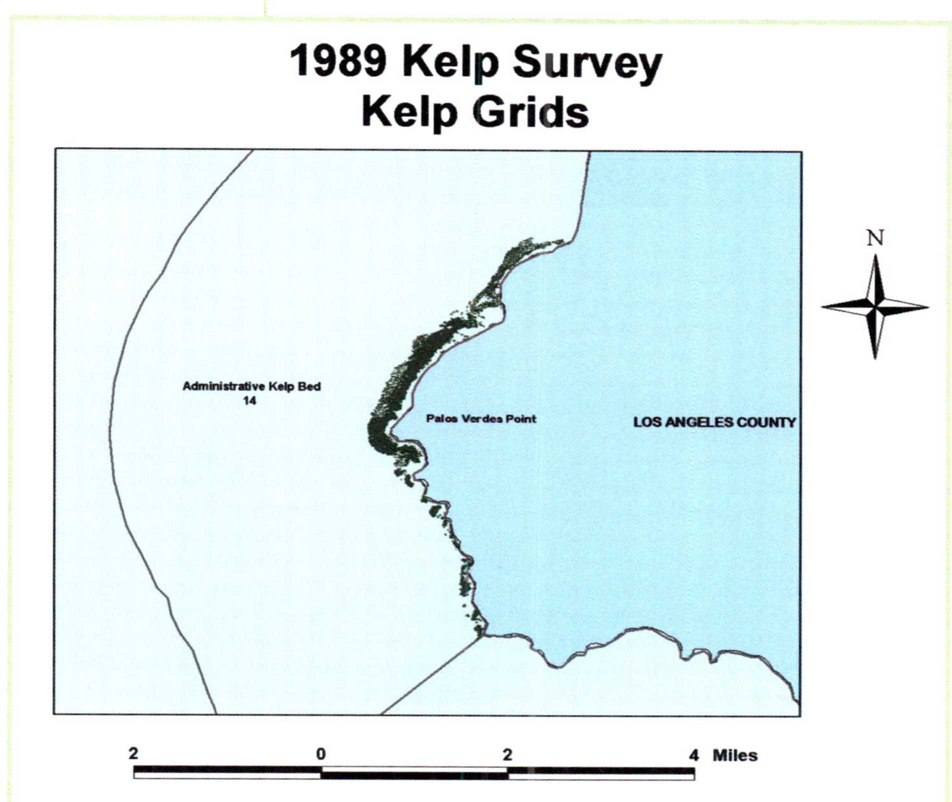

Eyes in the sky

As with past kelp surveys, the newest effort showed that accurate imagery was vital. Improving on past surveys, three sets of aerial photos were used this time to create the maps. Landsat satellite imagery was used to cross-reference and check the mapping from the large-format aerial photos. The decision to use large-format aerial photos for the new kelp bed survey came in the late summer and fall season, just before the annual cycle of winter storms began. Veisze and his DFG team used two private aerial photo firms and the DFG's own aerial photo unit to complete the job.

"I knew I had to throw a lot of resources at the project to ensure that it got done," Veisze said. "We were literally trying to stay ahead of the storms because they would come in and rip out the kelp we were trying to survey." From October 13 until December 22, 1999, the DFG team took 1,698 aerial photographs in 9-by-9-inch format of the kelp beds along the entire coast and the Channel Islands. A Garmin GPS 12XL global positioning system receiver determined the exact location of each photo. Chronic fog made it impossible to take pictures of a few miles of the Mendocino County coastline, the only area not included in the survey. The DFG team made contact prints of the 9-inch film and, in a money-saving move, scanned the prints instead of directly scanning the film.

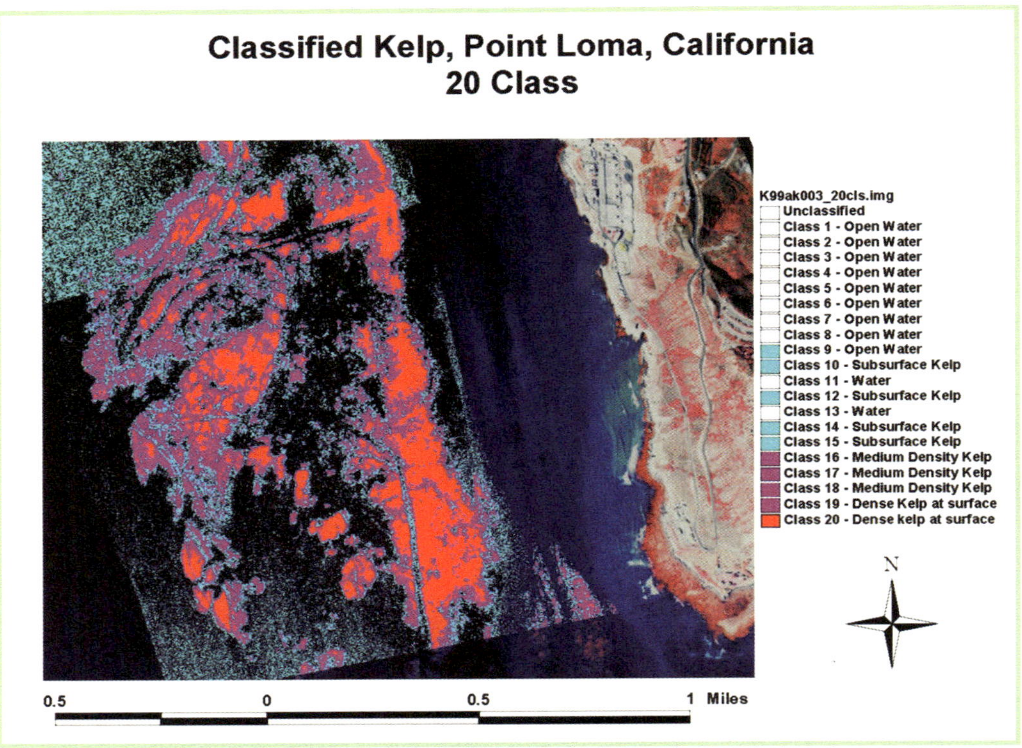

Figure 3.5 This map shows the large kelp beds off Point Loma in San Diego, California, divided into twenty separate classifications based on density and depth in the water.

Georeferencing made easy

The scanned aerial photos were georeferenced with the basemaps using two registration methods. Initially, photogrammetric processing with ERDAS IMAGINE OrthoBASE 8.4 software was used on single frames of aerial photos and the process employed camera parameters, camera orientation, and position. The process also corrected for any scale changes due to changes in terrain and elevation. An affine transformation was used for georeferencing the aerial photography in ArcView GIS 3.1 running the ArcView Image Analysis for ArcGIS extension version 1.

Veisze found that the ArcView Image Analysis extension sped up the process of georeferencing the aerial photos with the basemaps because it enabled real-time feedback on the fit of the registration. "You could pick a matching point on the image to base points on the map and fit the image to the base," he said.

After the aerial photos were georeferenced, the DFG team created separate ArcView shapefiles for each photo series. A polygon delineating the net coverage of each photo frame was screen-digitized over the photo display itself. The wire-frame indexes were prepared in addition to the initial photo-center-only spot indexes. Depending on the mapping requirements, simpler spot indexes or the more precise photo footprints were used.

Using ERDAS IMAGINE and ArcView Image Analysis software, Veisze and his team created an iterative, self-organizing clustering process in order to segment the images based on reflected light signatures. The reflected light signatures were then used to analyze kelp canopy and separate it into four classes ranked by density.

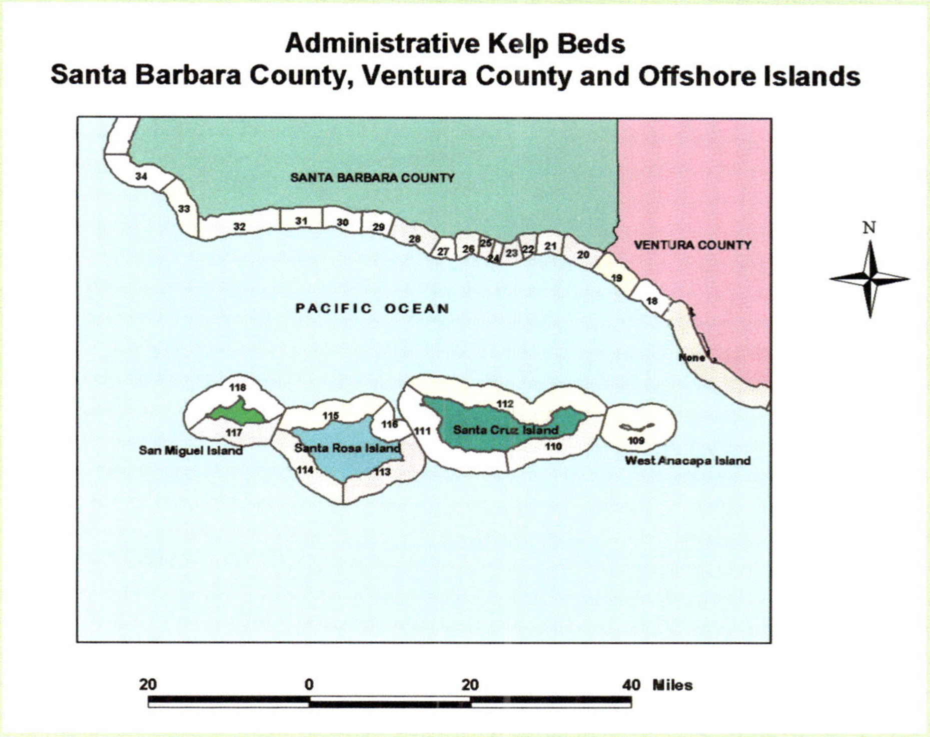

Figure 3.6 The administrative kelp beds stretch along most of the California coast and around many of the islands offshore. These kelp beds are along northern Ventura County and southern Santa Barbara County and the nearby islands.

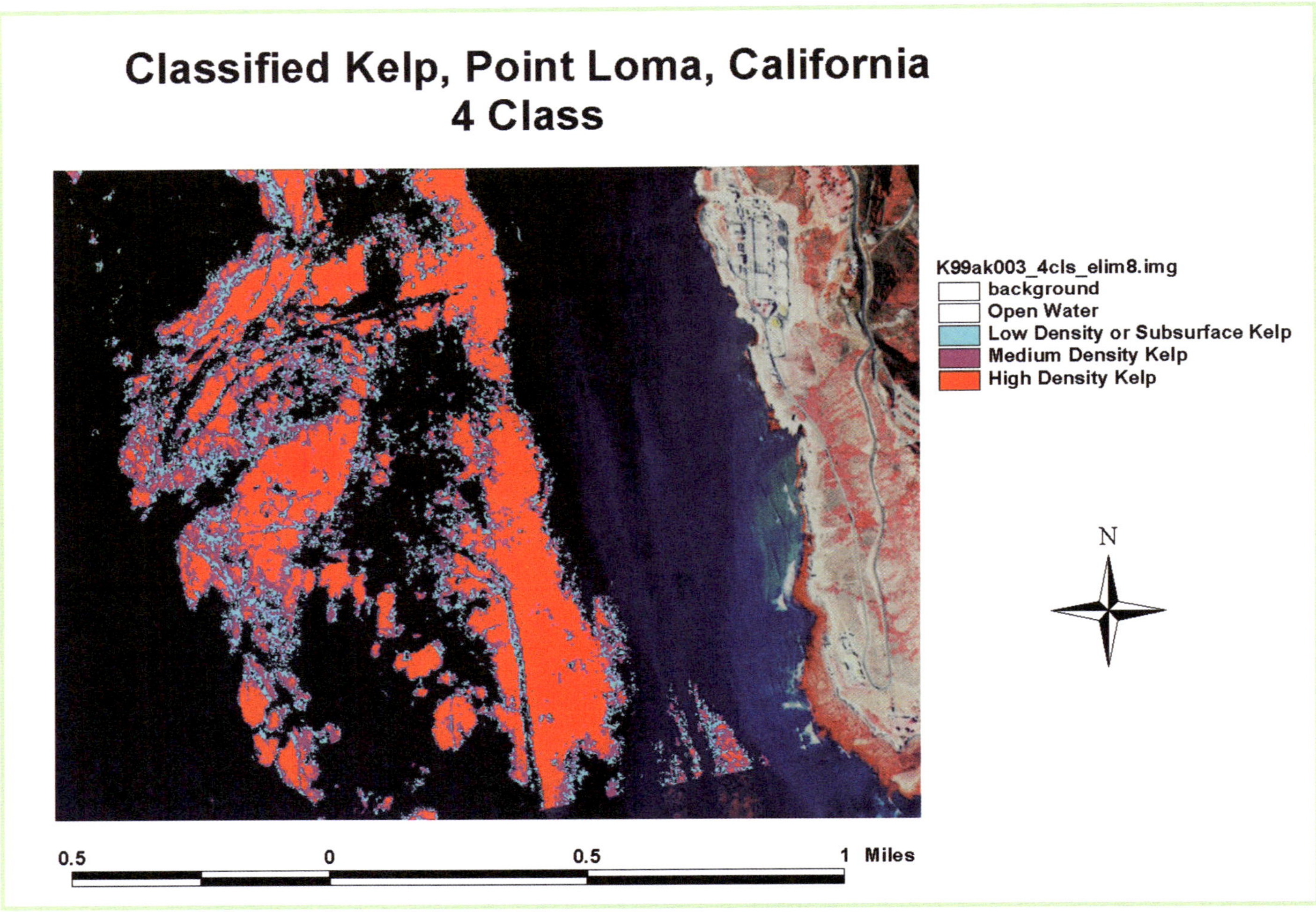

Figure 3.7 The areas of all classes of kelp (shown in figure 3.6) were summed in the reporting of kelp surface canopy by administrative kelp bed.

Calculating the kelp canopy's size

The DFG team also calculated the size of the kelp canopy. The data for the 1999 kelp survey was in raster format, and calculating area was a product of the number of pixels of kelp times each pixel's area with a scale of 4 square meters per pixel. The team calculated kelp area values within the GIS using ERDAS IMAGINE software. Calculating the area and density of each kelp bed administrative area also meant converting the statistics from meters to acres and miles.

Veisze and his team cross-checked the kelp maps created from the aerial photos with Landsat satellite imagery from the summer of 1999. Although the satellite imagery's 30-meter resolution was lower than that of the scanned 2-meter aerial photos, the kelp beds were still very distinct. To Veisze, the cross-checking proved effective because the Landsat imagery showed that some of the aerial photos taken in late November 1999 after a series of storms under-represented the size of kelp beds in certain areas. Veisze cautions against using the results of the more current DFG kelp surveys for comparison with the 1989 survey or with other past studies because of differences in methodology and technology used.

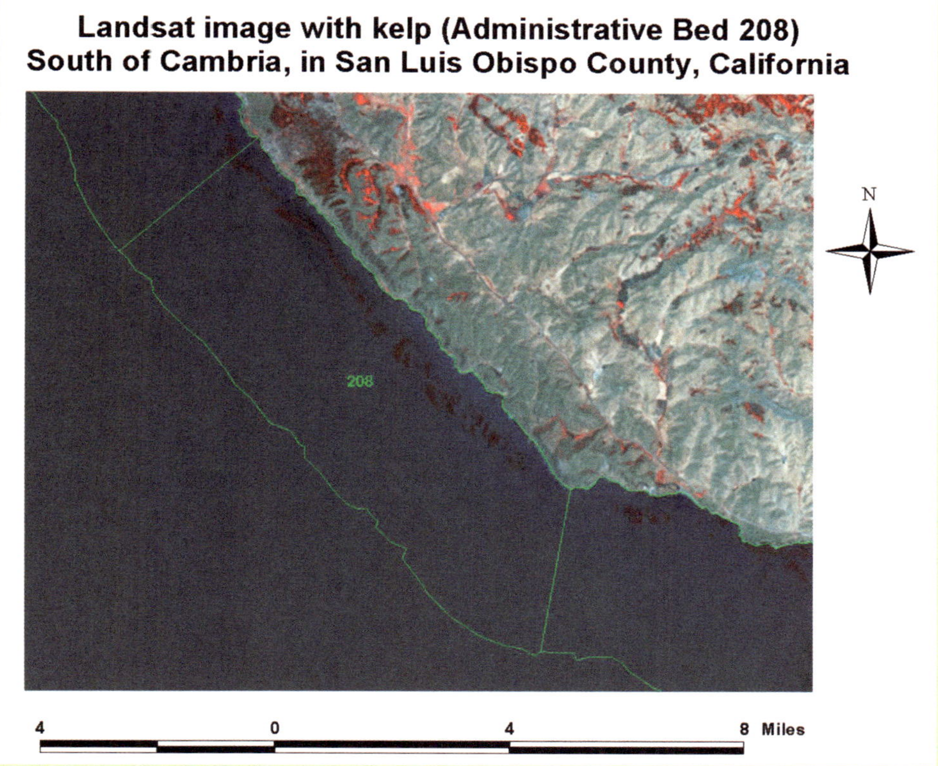

Figure 3.8 Landsat satellite imagery shows the kelp surface canopy off the coast of central California near the town of Cambria in San Luis Obispo County.

Technological improvements

Dennis Bedford, associate marine biologist with the California Department of Fish and Game, has overseen the kelp survey project since 2002. According to Bedford, advanced technology has improved and simplified California's kelp surveys. Since the 1999 kelp mapping, the DFG has acquired a SpecTerra digital multispectral video system.

SpecTerra uses the same spectral imaging techniques used for the 1999 survey's aerial photo images. The same basemaps used for the 1999 survey were used for georeferencing the 2002 survey. Processing digital video data differs from processing aerial photos. MicroImages TNTmips software was used to georeference the digital images instead of ERDAS.

Bedford found that video technology has enabled the DFG to conduct kelp surveys annually instead of every decade or so. Using SpecTerra, the DFG surveyed the kelp beds each year since 2002, enabling the DFG to build a time sequence of kelp surveys. The time sequences can help detect changes in the kelp canopy from year to year. SpecTerra also allowed the DFG to bring the kelp bed survey work in-house while enabling the agency to monitor other coastal resources and, if necessary, make multiple kelp surveys of specific areas within a given year.

Commercially available high-resolution satellite imagery is another technological advance that has emerged since the 1999 survey. Landsat, which covers the entire coast with nine images—compared with hundreds of aerial photos—would be well suited for quality control and cross-referencing the video images from the SpecTerra system. Due to budget constraints, however, the acquisition of expensive high-resolution satellite imagery is unlikely, according to Bedford.

3.9a

Source: Dennis Bedford, Associate Biologist, Coordinator, California Department of Fish and Game, Kelp Monitoring and Management Program; California Resources Agency © State of California.

Source: Dennis Bedforc, Associate Biologist, Coordinator, California Department of Fish and Game. Kelp Monitoring and Management Program; California Resources Agency © State of California.

Source: Dennis Bedford, Associate Biologist, Coordinator, California Department of Fish and Game. Kelp Monitoring and Management Program; California Resources Agency © State of California.

Figures 3.9a, 3.9b, and 3.9c Kelp strongly reflects the infrared portion of the spectrum, and this type of analysis helps show the true extent of a kelp bed, since portions may be hidden under a few feet of water and hard to see in a full-color photograph. These images are not the actual photographs, but are an analysis based on the photographs.

More than commercial value

Bedford noted that monitoring the health of California's kelp has value beyond its immediate commercial applications. "Harvesting kelp in California is really minor compared to the total stock of standing kelp," he said. California's kelp forests, like their terrestrial counterparts, are a unique habitat. The continued health and survival of many species depend upon the kelp beds. The kelp-bed GIS may play a larger role in conserving California's coastal zone resources. The GIS may also help the state create new underwater preserves that include kelp forests, which would help revamp the management of California's fisheries.

GIS helps improve the effectiveness of marine reserves. Protecting certain key species might promote kelp growth. Sea urchins feed on kelp and, in the absence of natural predators, urchins overpopulate, leading to areas devoid of kelp. Lobsters and fish such as the sheephead wrasse feed on sea urchins. Restricting the harvest of urchin predators in specific areas could encourage the growth of some kelp beds, thus helping to preserve this vital habitat for future generations.

"One purpose of the GIS is longer term, to help manage fish species," Bedford said. "Kelp canopy plays a vital role in the survival of juvenile fishes. In some cases we are finding a strong correlation between canopy cover in any year and the number of adult fish which would have survived as juveniles in that year. The implications for future management of some fish species are tremendous."

3.10b

Photo by Robert Scally

Photo by Robert Scally

Figures 3.10a and 3.10b Lobsters and sea urchins are two common residents of kelp forests. Pacific spiny lobsters, like the one shown here, are predators of sea urchins, which can destroy kelp forests if they become too plentiful. Surveys of the California kelp beds can help spot problems in kelp beds and detect lobster over-fishing and other problems.

References

Kilgore, Alan, Mark Lampinen, Paul Veisze. 2001. Building a California kelp database using GIS. Presented at ESRI User Conference, San Diego, Calif. July 9–13
gis.esri.com/library/userconf/proc01/professional/papers/pap900/p900.htm

Wetlands 4

Drawing the line between wet and dry along Lake Ontario

The Rochester Embayment Area of Concern stretches along the shore of Lake Ontario in Monroe County, New York.

Data provided by 2004 ESRI Data & Maps.

Lake Ontario's shoreline near Rochester, New York, is a wetland wonderland under threat. A series of small, shallow ponds, bays, vernal pool clusters, and marshes dot the lake's edge and the adjacent uplands of Monroe County, New York. This area shares challenges with many other wetlands in the continental United States. Monroe County's once expansive wetlands were altered first in the nineteenth and early twentieth centuries by agriculture and drainage projects. Early in the twenty-first century, the county's wetlands face new development pressures and are becoming golf courses, industrial parks, and housing tracts. GIS is playing a key role in assessing and managing these valuable wetlands.

Figure 4.1 Narrow, well-defined wetlands in Durand-Eastman Park near Rochester, New York, are among the many types found within the Rochester Embayment. In this aerial view, colorful fall foliage helps define wetland and upland boundaries.

Image courtesy of Pictometry International.

America's wetlands—its swamps, bogs, marshes, and estuaries—are disappearing at a rate of about 117,000 acres annually. Less than 100 million acres remain of the approximately 215 million acres of wetlands that existed in the continental United States two hundred years ago, according to the National Audubon Society. Wetlands once were considered wastelands and swamps, nothing more than natural impediments blocking commercial progress. It is now understood that wetlands not only provide habitat for a variety of plants and animals, they also function as a kind of natural flood control by absorbing excess water during rainy seasons or snow melts. Wetlands even act as pollution filters, absorbing toxins that are in water and bonded to sediments.

Image courtesy of Pictometry International.

Figure 4.2 The same wetlands shown in figure 4.1 are depicted here in summer, with state wetland boundaries superimposed. The golf course in the lower portion of the photo was formerly a wet and, according to state maps. Water still stands in one portion of the golf course, a remnant of the site's natural past.

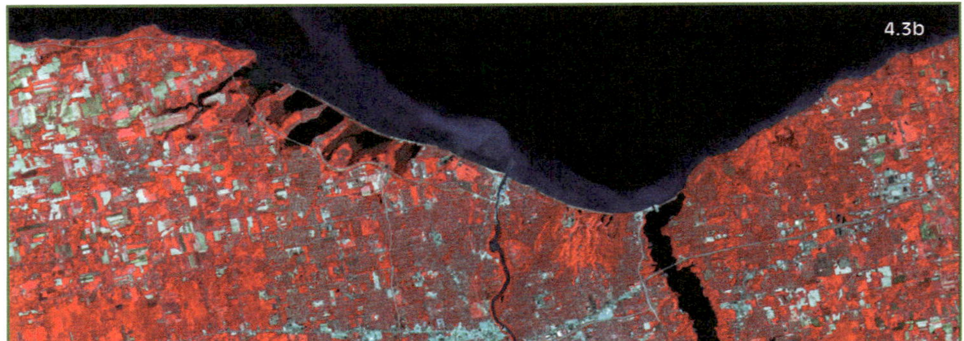

Figures 4.3a, 4.3b, and 4.3c Landsat satellite imagery composite views of the Rochester Embayment in true color (bands 1, 2, and 3), and two false-color composites (bands 2, 3, and 4, and bands 3, 4, and 5) illustrate the mixed land uses and land covers, such as residential development and agriculture, in the embayment drainage that impact wetland size and health.

Maps by Dr. Karl Korfmacher, Rochester Institute of Technology. Landsat imaging courtesy of the USGS.

Preserving upstate New York's wetlands

Dr. Karl Korfmacher, an environmental science professor at the Rochester Institute of Technology (RIT), is using GIS to determine how development, agriculture, and other byproducts of human activity have affected the Rochester Embayment Area of Concern (AOC). In 2003, Korfmacher conducted a pilot program, underwritten by the small grants program of the New York State Great Lakes Protection Fund, to study six wetlands in detail within the Rochester Embayment AOC.

According to Korfmacher, most of Monroe County was probably once wetlands. Successive waves of agricultural development, industrialization, and urbanization have claimed much of the area's wetlands. Rural areas were drained and crisscrossed with irrigation ditches for vegetable farms and fruit tree orchards. Much of this former farmland is now becoming suburbs, further impacting adjacent, remaining wetlands.

"Our project was designed to figure out how big the wetlands currently are and to take field surveys through them to determine the quality of the wetlands by looking at plant biodiversity and the number of invasive species," Korfmacher said. "Are we losing extent because it's being

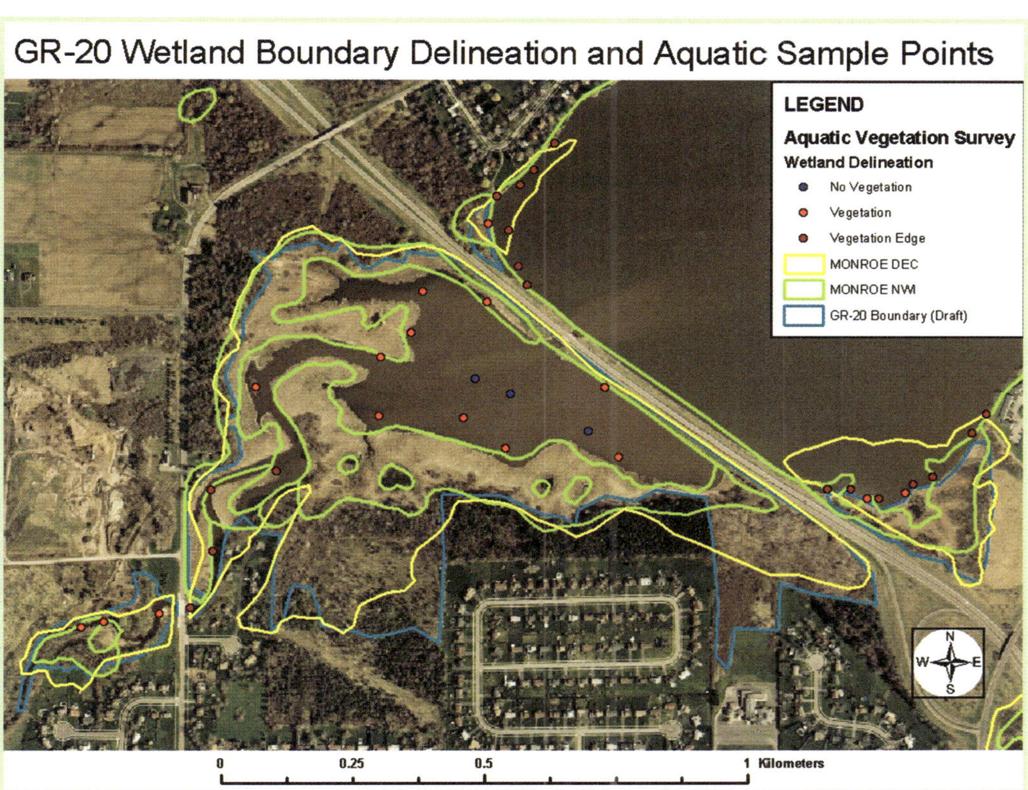

Map by Dr. Karl Korfmacher, Rochester Institute of Technology. Orthoimagery from the New York GIS Clearinghouse.

Figure 4.4 Three different wetland boundary delineations for the Long Pond wetland complex (GR-20) showing field survey aquatic sample points and the proximity of the wetlands to a residential neighborhood built between 1983 and 1996.

eaten up by urban development or agriculture? How much of an impact is there?" Korfmacher's questions are especially relevant to certain land-use restrictions around the Rochester Embayment AOC.

The pace of American wetland destruction slowed in the 1990s after the federal Clean Water Act and its amendments were enacted along with state and local laws protecting wetlands. Today, development in and around wetlands is often restricted. But it has not halted. Despite new laws and improved protections, residential and commercial developments on small parcels are still erasing wetlands a few acres at a time.

One of the most prominent conservation efforts came in 1972 with the Great Lakes Water Quality Agreement between the United States and Canada. The agreement focuses on efforts to "restore and maintain the chemical, physical, and biological integrity of the waters of the Great Lakes Ecosystem." The agreement led to a series of resource restrictions—known as use impairments—that helped define forty-two areas of concern in the Great Lakes.

The Rochester Embayment AOC covers the drainage areas along Lake Ontario near Rochester, between Bogus Point near the town of Greece, and Nine-Mile Point near the town of Webster, and up to the Lower Falls of the Genesee River (approximately 6 miles).

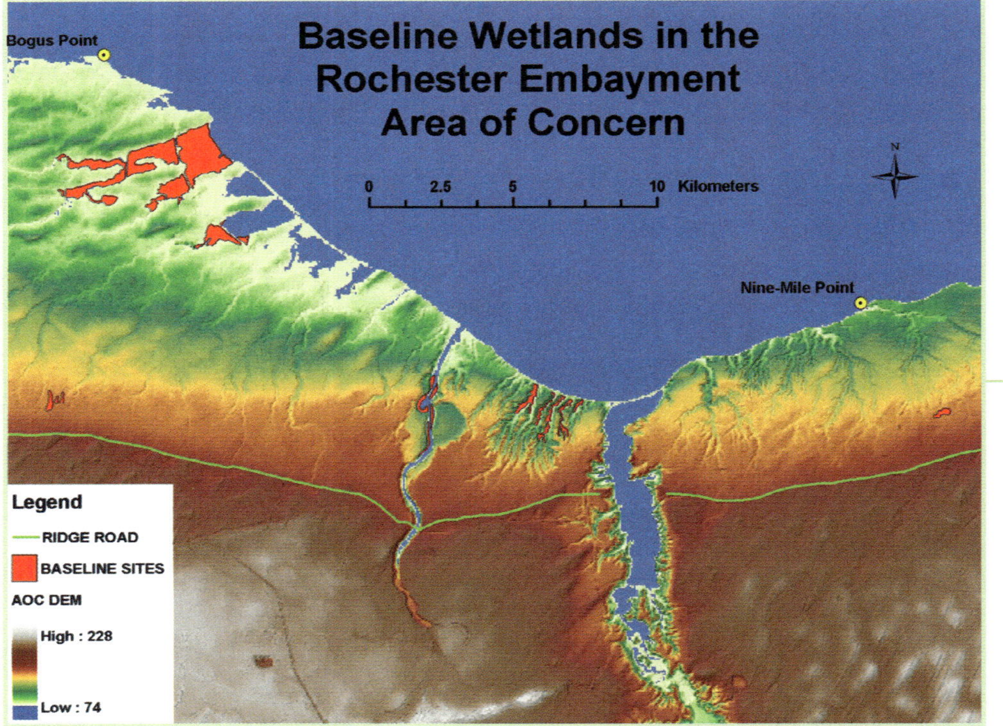

Map by Dr. Karl Korfmacher, Rochester Institute of Technology. Orthoimagery from the New York GIS Clearinghouse.

Figure 4.5 Baseline wetlands boundaries in the Rochester Embayment AOC overlaid on a digital elevation model (DEM) showing drainage patterns and elevation differences within the study area. Ridge Road (green line) formed the southern boundary of the study.

The Rochester Embayment AOC's drainage area covers more than 4,828 square miles. Because of its AOC designation, a remedial action plan was prepared and implemented for the Rochester Embayment. The plan identifies fourteen property-use impairments and criteria for addressing and removing these impairments. Use Impairment 14 concerns "loss of fish and wildlife habitat," and wetlands play a critical role in this issue.

The following are the criteria for removing Use Impairment 14:

- No net loss of acreage and quality of federal or state designated wetlands using a baseline year for comparisons
- No net loss of the 50-foot-wide buffer strip of trees and shrubs on both sides of streams classified by the New York State Department of Environmental Conservation using a baseline year for comparisons

Measuring change

The ongoing GIS project is important to Monroe County's wetlands because of pressure to develop near and along the shoreline. One of the primary use impairments in the Rochester Embayment AOC requires that no net loss of fish and wildlife habitat take place due to wetland destruction caused by deforestation, agriculture, or urban or suburban development. "For each of the six bayside wetlands, we're trying to identify what are the major pressures on that system and whether or not [the wetlands] are getting larger or smaller as a result of those pressures," Korfmacher said.

Beginning in 2003, Korfmacher and his students began using GIS to help define and survey the embayment wetlands. Their mission was to create accurate maps of the Rochester AOC wetlands, establishing a baseline year of 2003 to measure changes against, and to show subsequent changes in the wetlands. Basemaps for the project were created from digital orthophoto quarter quadrangle (DOQQ) photography with 1-foot to 3-foot resolution and USGS topographic maps.

Korfmacher's project draws on a variety of data sources because he is using both the New York Department of Environmental Conservation wetland delineation methods, which focus primarily on size and vegetation, and the U.S. Army Corps of Engineers' method for defining wetlands. The Corps of Engineers' wetlands protocol requires that wetlands have the right types of vegetation, hyrodology, and soil, each with its own data layer within Korfmacher's GIS analysis. The study also makes use of the Monroe County Soil Survey Geographic (SSURGO) database—part of the federal soil database—to help identify wetlands.

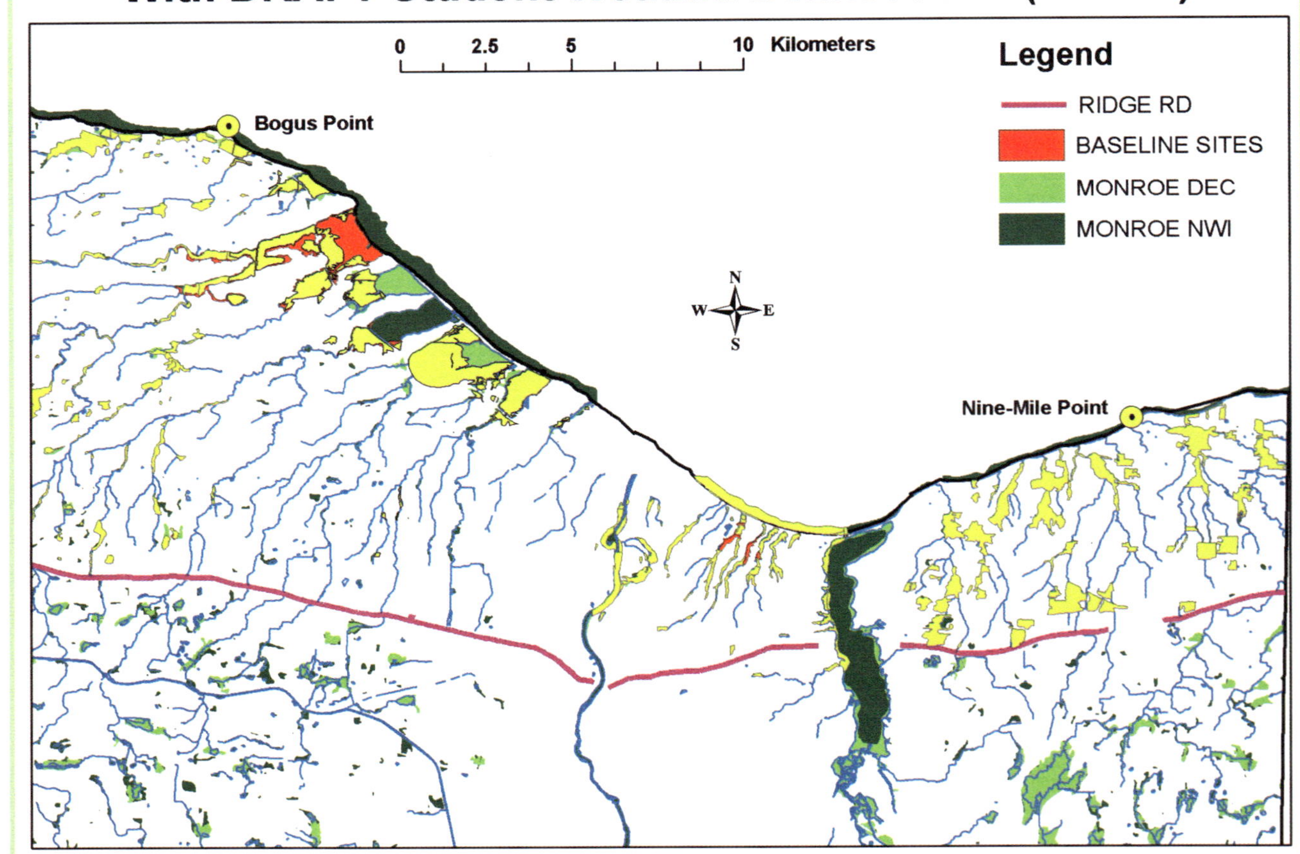

Map by Dr. Karl Korfmacher, Rochester Institute of Technology. Orthoimagery from the New York GIS Clearinghouse.

Figure 4.6 Dr. Karl Korfmacher's students used ArcGIS, Landsat imagery, digital orthophoto quarter quadrangles (DOQQs), and the results of their field surveys to produce this draft basemap that delineates known and potential wetlands in the Rochester Embayment AOC.

Federal and state government teams conducted wetland surveys of Monroe County in the 1970s and 1980s. These surveys may have been accurate when originally conducted, Korfmacher points out. But land-cover and land-use changes have undoubtedly altered the delineated wetland boundaries in the years since the previous surveys. However, Korfmacher used the surveys as starting points for his research, since they represent the official wetland areas.

"The other key part of the study was looking at how wetlands change over time," Korfmacher said. Thanks in part to the proximity of the Rochester headquarters of Eastman Kodak, the area has a long record of aerial photography, a factor that makes time sequences very effective. Aerial photos of the wetlands from the 1930s through the 1990s that have been georeferenced and photo-mosaicked for the Monroe County government became a key resource for this project. Korfmacher used Landsat images to create seasonal maps and time sequences.

"What we're trying to do is show that you can use these older aerial images, these older Landsat images in time series to show if there's been increases or decreases in the wetlands," Korfmacher said. "Not all land-cover change necessarily destroys wetlands."

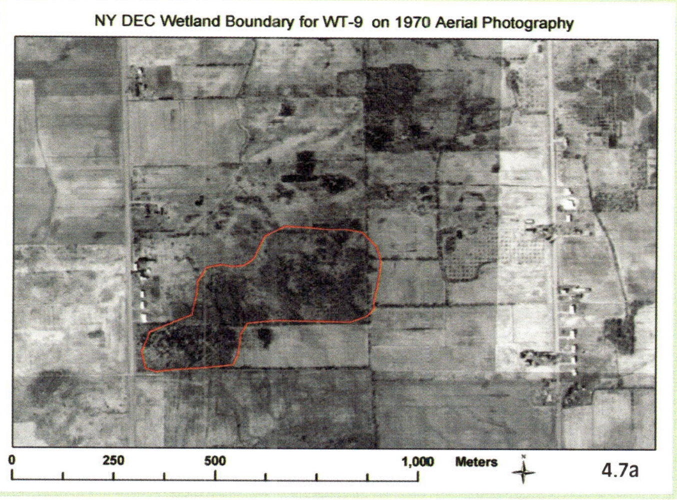

Mosaicked, georeferenced aerial photos provided by the GIS Division of the Monroe County Department of Environmental Services, GIS Service Division.

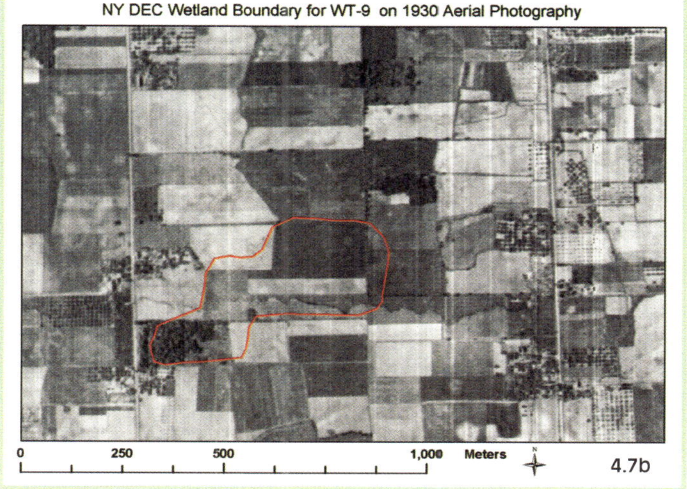

Mosaicked, georeferenced aerial photos provided by the GIS Division of the Monroe County Department of Environmental Services, GIS Service Division.

Figures 4.7a and 4.7b Two aerial photos illustrate wetland changes in New York State DEC wetland delineation for the Webster wetland complex (WT-9) between 1930 and 1970.

Ground truthing

Korfmacher and his students didn't sit behind their computers relying on data created by others. They pulled on their boots, sprayed on some DEET insect repellent, and spent six weeks in the field gathering georeferenced vegetation samples and ground truthing the basemaps, aerial photos, and Landsat images. Ground truthing is the process of conducting ground surveys to confirm details and features shown on maps, aerial photos, or satellite images. In this case, Korfmacher and his students hiked through their study areas to see if reality matched what the maps and photos depicted.

They recorded data on the herbaceous layers under 1 meter high, a shrub layer up to 3 meters high, and the tree layer over 3 meters at each sample point. They recorded hydrology characteristics and used soil samples they took at each point to assess hydric properties. Korfmacher and his students recorded aquatic data for sites where permanent water was present. They used data for all plant species to help determine biodiversity and the impact of invasive species, in addition to delineating wetland or upland classifications.

Korfmacher and his students used GPS receivers linked to Compaq Pocket PCs running ArcPad software to record data point locations. ArcPad software allowed the field teams to see their positions on the basemaps in real time, aiding in the wetland boundary delineation. Data entered onto the Pocket PCs was later uploaded to desktop computers. Time, weather, and money limited the scope of the fieldwork. As a result, Korfmacher's team focused primarily on three of the six wetlands included in the study, although all six wetlands were surveyed. The primary wetlands studied were those contained within Durand-Eastman Park and Braddock Bay on Lake Ontario, as well as a wetland located on farmland near the town of Webster.

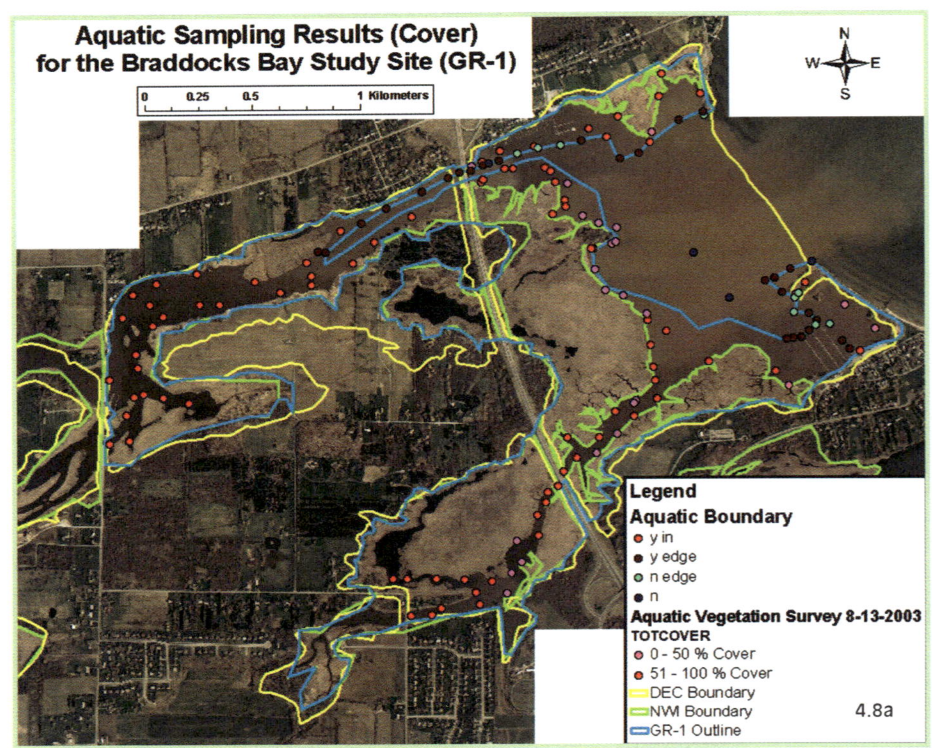

Map by Dr. Karl Korfmacher, Rochester Institute of Technology. Orthoimagery from the New York GIS Clearinghouse.

Figures 4.8a, 4.8b, and 4.8c Three maps showing boundary and field survey data points along the Braddock's Bay wetland complex (GR-1). The maps illustrate the diversity of wetland vegetation and percentage of vegetation cover within the sampling areas.

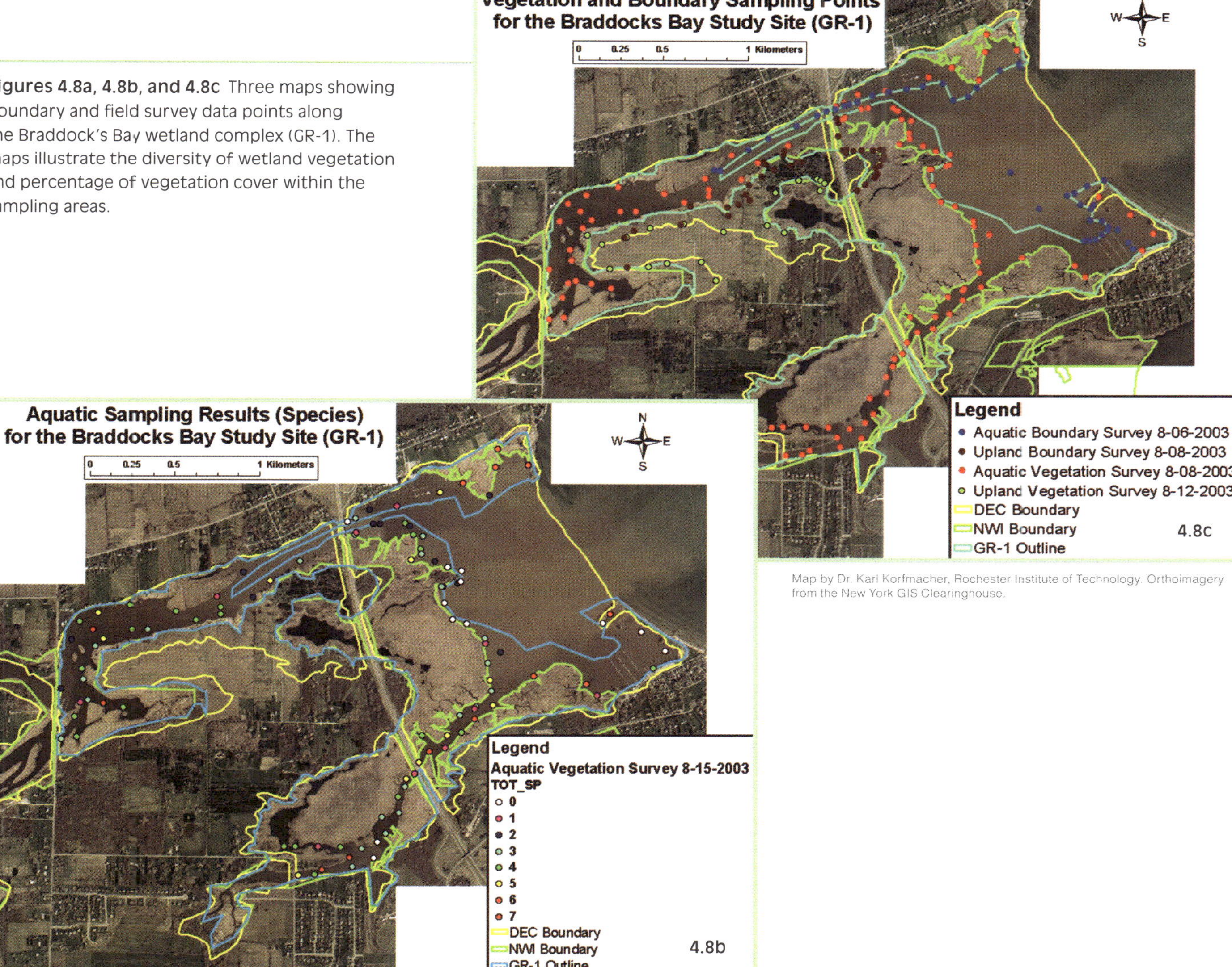

Map by Dr. Karl Korfmacher, Rochester Institute of Technology. Orthoimagery from the New York GIS Clearinghouse.

Map by Dr. Karl Korfmacher, Rochester Institute of Technology. Orthoimagery from the New York GIS Clearinghouse.

Wetlands of the rich and famous

One area that the team studied closely was the Durand-Eastman Park wetlands. Dr. Harry S. Durand, a prominent local physician practicing in the 1880s and 1890s, and Eastman Kodak Company founder George Eastman donated the 965-acre property to the city in 1909. "Durand-Eastman Park looks like a miniature version of the Finger Lakes," Korfmacher said, referring to the famous lakes in central New York.

During the spring and summer, Durand-Eastman Park is a lush, green, bird haven. The park's more than 5,000 feet of shorefront attracts sunbathers. Bird watchers report seeing migrating warblers, vireos, and flycatchers in the park during the spring and fall. In the winter, cross-country skiers flock to the park along with wintering birds such as scaup, crossbills, and finches.

Nearby Braddock Bay, which Korfmacher's team also studied, is a key migration point for songbirds and especially raptors. Korfmacher chose to study the park because its wetlands are well defined topographically. "Those wetlands [in the park] are basically confined by the very steep slopes around them," he said. The park is also a good example of the pressures facing wetlands throughout the continental United States. Durand-Eastman's park and its wetlands have been subject to the siren call of development.

In the early twentieth century, the former Durand and Eastman estates became public land. A portion of the wetlands became the eighteen-hole Durand-Eastman Golf Course, and the Frank E. Van Lare Wastewater Treatment Plant was constructed adjacent to the park's eastern edge. Suburban Rochester is immediately to the west of the wetlands.

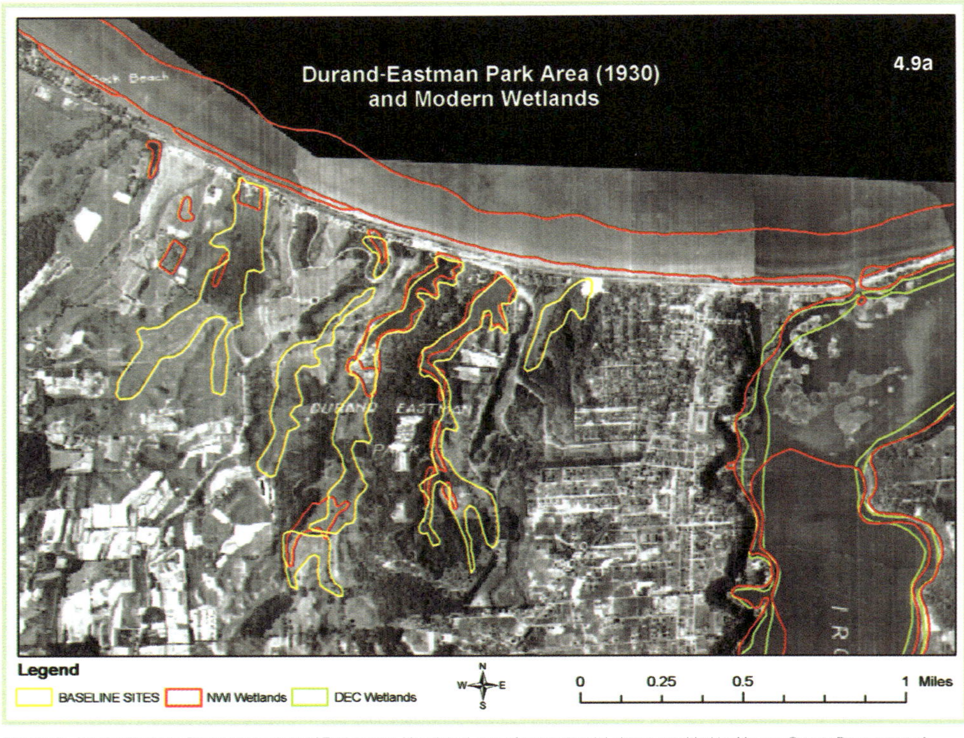

Map by Dr. Karl Korfmacher, Rochester Institute of Technology. Mosaicked, georeferenced aerial photos provided by Monroe County Department of Environmental Services, GIS Service Division.

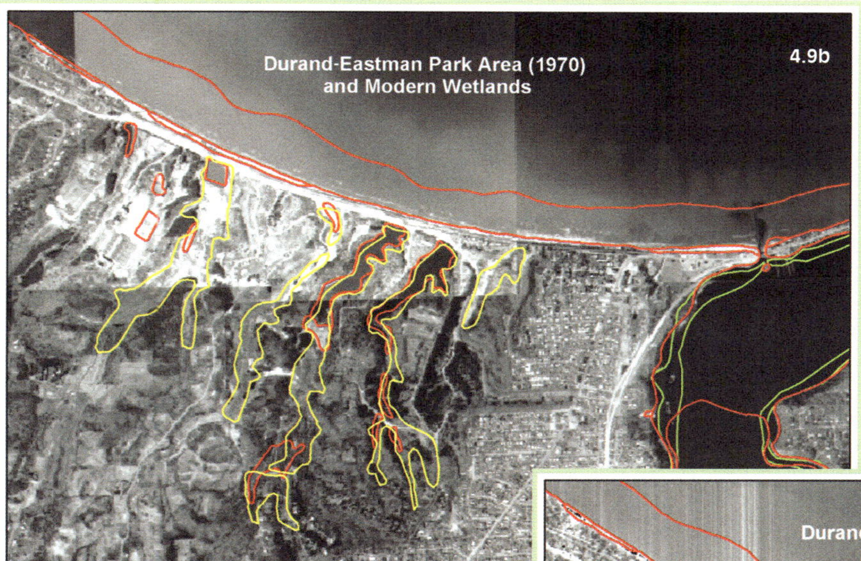

Map by Dr. Karl Korfmacher, Rochester Institute of Technology. Mosaicked, georeferenced aerial photos provided by Monroe County Department of Environmental Services, GIS Service Division.

Figures 4.9a, 4.9b, and 4.9c This fifty-year time sequence shows land-use and land-cover changes in the area surrounding the Durand-Eastman Park wetlands area. The aerial photos show the area in 1930, 1970, and 1980. Note the golf course in the wetland polygons on the left side of the 1980 photo.

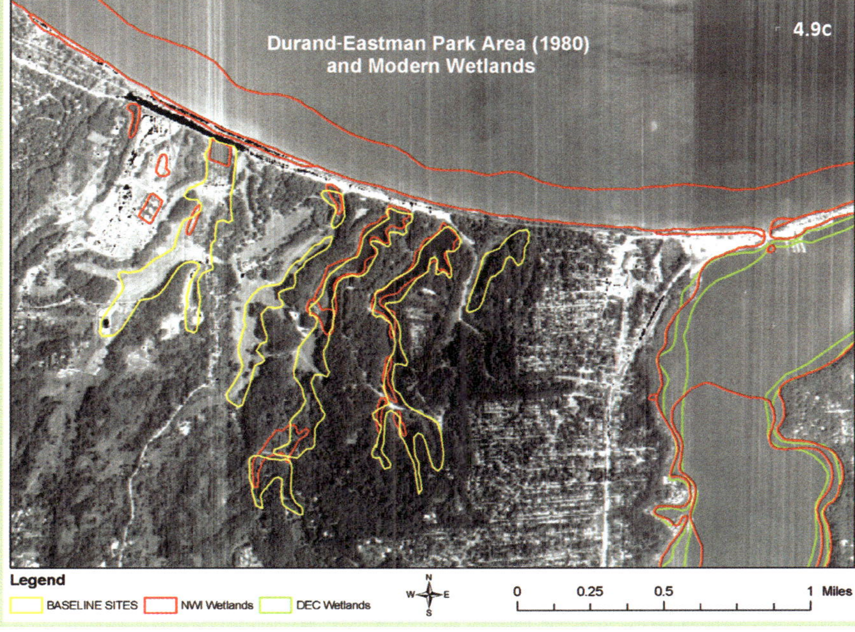

Map by Dr. Karl Korfmacher, Rochester Institute of Technology. Mosaicked, georeferenced aerial photos provided by Monroe County Department of Environmental Services, GIS Service Division.

An improved wetlands baseline

Korfmacher's team examined the park's well-defined wetlands using Landsat false-color imagery. False-color imagery uses selected colors instead of shades of gray or natural color, to dramatically depict different levels of brightness and highlight very small differences in features in an image. The team processed the Landsat images using IDRISI software and then imported them into ArcMap software for viewing. IDRISI software is raster-based GIS and image-processing software for analyzing satellite or aerial images, among other tasks.

Using aerial photos combined with the basemaps, Korfmacher then added layers showing the wetland boundaries as depicted by the Federal National Wetlands Inventory survey, the New York Department of Environmental Conservation survey, and the baselines established by the project's research. Placing these same layers over aerial photos of the park from 1930, 1950, 1960, 1970, 1980, and 2002 created time sequences.

Incorporating data from federal and state sources, aerial and satellite images, and combined with fresh field research, GIS is helping paint a more accurate picture of the wetlands in Monroe County that could have implications for both public policy and the environment. "If those federal and state boundaries are wrong, either initially mismapped or as a result of changes over time, in many cases it would be very difficult to show a net increase or a net loss of wetlands within the area of concern," Korfmacher said.

As they began analyzing the spatial data and creating maps, Korfmacher and his students quickly found major differences between their own data and earlier federal

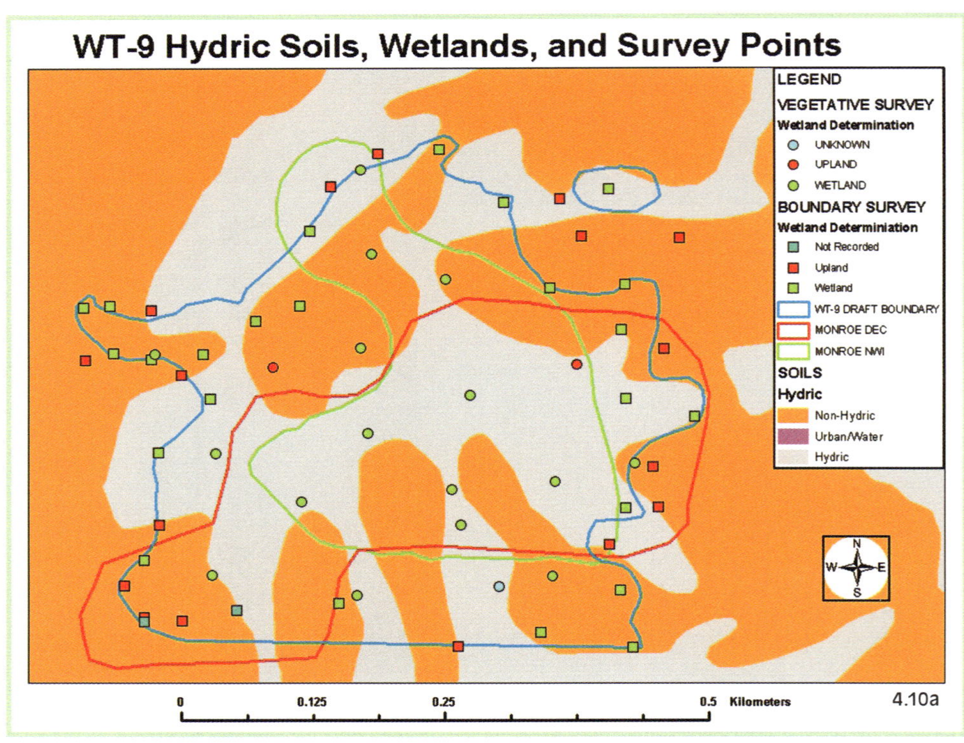

Map by Dr. Karl Korfmacher, Rochester Institute of Technology. Orthoimagery from the New York GIS Clearinghouse.

and state wetlands surveys. The previous surveys showed some wetlands as being smaller than the 2003 survey suggests, while others were much larger than Korfmacher's team found. The new research also found that urban development does not necessarily reduce wetlands. In one case, runoff from recently developed land was actually feeding adjacent wetlands. What is unclear is how that runoff will ultimately impact the wetlands' health.

Through field research and GIS analysis, Korfmacher's study also discovered several previously undocumented wetlands. Korfmacher has applied for more research funding. The project is already producing baseline data that better reflects current conditions and extents than the previous federal and state surveys, which are being used as the legal guidelines for setting wetland boundaries and imposing development restrictions. Thanks to the research and GIS analysis conducted by Korfmacher's team, the line between wetland and dry land is becoming more clearly and accurately drawn and potentially better managed.

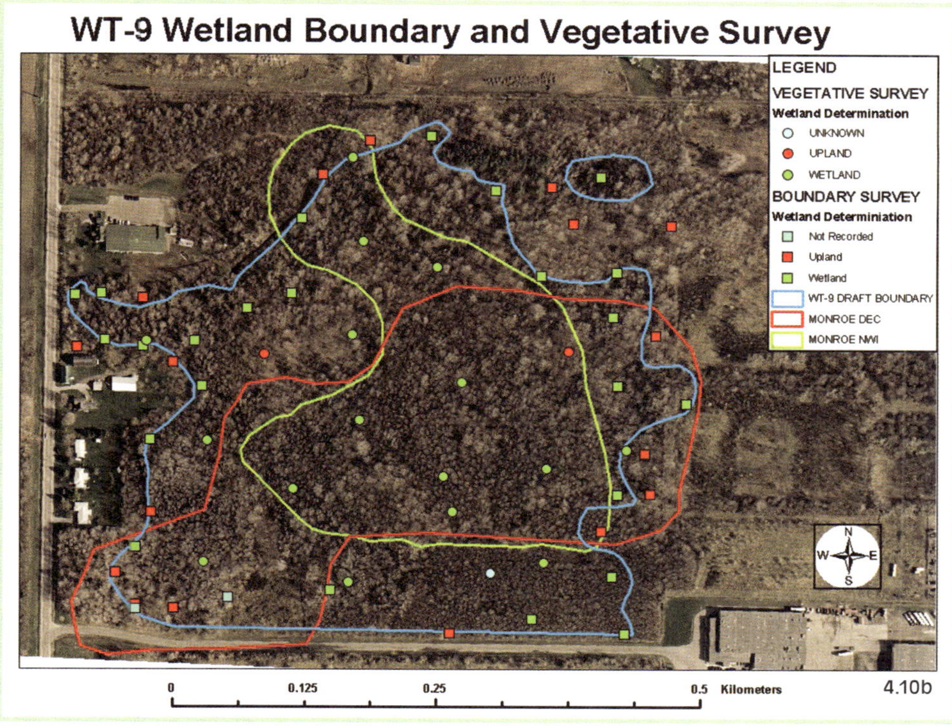

Map by Dr. Karl Korfmacher, Rochester Institute of Technology. Orthoimagery from the New York GIS Clearinghouse.

Figures 4.10a and 4.10b Two maps showing results of survey and mapping work combined with DOQQs and data from the Natural Resources Conservations Service's national soils database. The soils data is classified for hydric characteristics at the Webster wetland complex (WT-9). Hydric soil maps helped refine the search for missed wetland areas and interpreted wetland boundaries from aerial photos, although soils are prone to change as well.

GIS helps preserve Japan's Kushiro Wetland

The 183-square-kilometer (71-square-mile) Kushiro Wetland in Hokkaido prefecture in northeastern Japan is home to a variety of wildlife, but none more notable than the Japanese crane. Strong and beautiful, cranes hold a special place in Japan's folk culture. The birds even adorn Japanese currency. The very symbol of the nation, Japanese cranes mate for life and are extremely loyal to their partner. Yet, by the early 1920s, Japanese cranes were nearly extinct. A few cranes discovered in the Kushiro Wetland in 1924 spawned a movement to protect them.

In the 1990s, encroaching agriculture and development threatened to destroy the important wetland. While the cranes may be the marsh's most famous residents, they aren't alone. Several endangered species, including Blakiston's fish owl and the Siberian salamander, also live in the Kushiro Wetland. More than six hundred plant species, twenty-six mammal species, four kinds of amphibians, five reptile varieties, thirty-five kinds of fish, and 1,150 types of insects call the wetland home.

In 2002, the Japanese Ministry of the Environment began a restoration project in the Kushiro Wetland in cooperation with private organizations, local governments, and related ministries and agencies. The Kushiro approach to preservation resulted in the Kushiro Wetland Natural Environment Information Map, a Web-based GIS database.

Figure 4.11 Japanese cranes, also known as *tancho,* were once thought to be extinct. Several cranes were found living in the Kushiro Wetland in 1924, sparking the beginning of an environmental preservation movement in Japan.

Hokkaido
Kushiro Wetland
JAPAN
Pacific Ocean
Data provided by 2004 ESRI Data & Maps

Kushiro Wetland in Northeast Hokkaido, Japan.

GIS provides communication tools

GIS helps the public learn about the restoration efforts and promotes data sharing among those working on the restoration, according to Dr. Masami Kaneko, associate professor of regional environmental studies at Rakuno Gakuen University. "There is no doubt that visual information is always easier to interpret than text or numbers," he said. "This can help the public to understand without specialized knowledge. Therefore, we can show the current situations and problematic points intuitively by using maps as part of the public education."

Kaneko is combining separate maps, databases, scientific studies, and aerial photos of the Kushiro Wetland into a single Web-based GIS database that is accessible to both researchers and the public. "Due to the characteristics of the wetland, it is necessary to understand not only the individual components of the wetland, but also the entire drainage basin in order to plan the conservation and restoration project efficiently," Kaneko explained. "As GIS has the merit of presenting an overall picture efficiently, it was chosen as a tool to aggregate scientific data systematically related to the wetlands."

Building a Web-based GIS

Kaneko spent two years building the Kushiro Environment Information Database, initially collecting basemaps that governmental agencies, such as the Geographical Survey Institute of Japan, had converted to GIS databases. He then assembled the limited chronological and detailed data.

The project's next stage was building up layers of detailed data, such as wetland areas, stream lines, vegetation, and farmland, delineated on topographical maps

Source: Masami Kaneko.

Figure 4.12 The Metadata Explorer browser application serves as both a search tool and a metadata warehouse.

and orthorectified aerial photos. Landsat, SPOT, and IKONOS satellite images of different time periods provided more chronological data. Converting documents to an Adobe Acrobat PDF file allowed for publication within a searchable database on the Web site.

The Kushiro Wetland Nature Restoration Project is a model for ecological restoration and preservation in Japan. The Kushiro restoration process cycle of "plan, do, check, and action" is based on GIS databases that will help conserve and restore other wetlands in Japan. "The development of the database is an ongoing project," Kaneko said. "In addition, we would like to encourage data and information sharing about wetland, which can contribute to wetland study and environmental education."

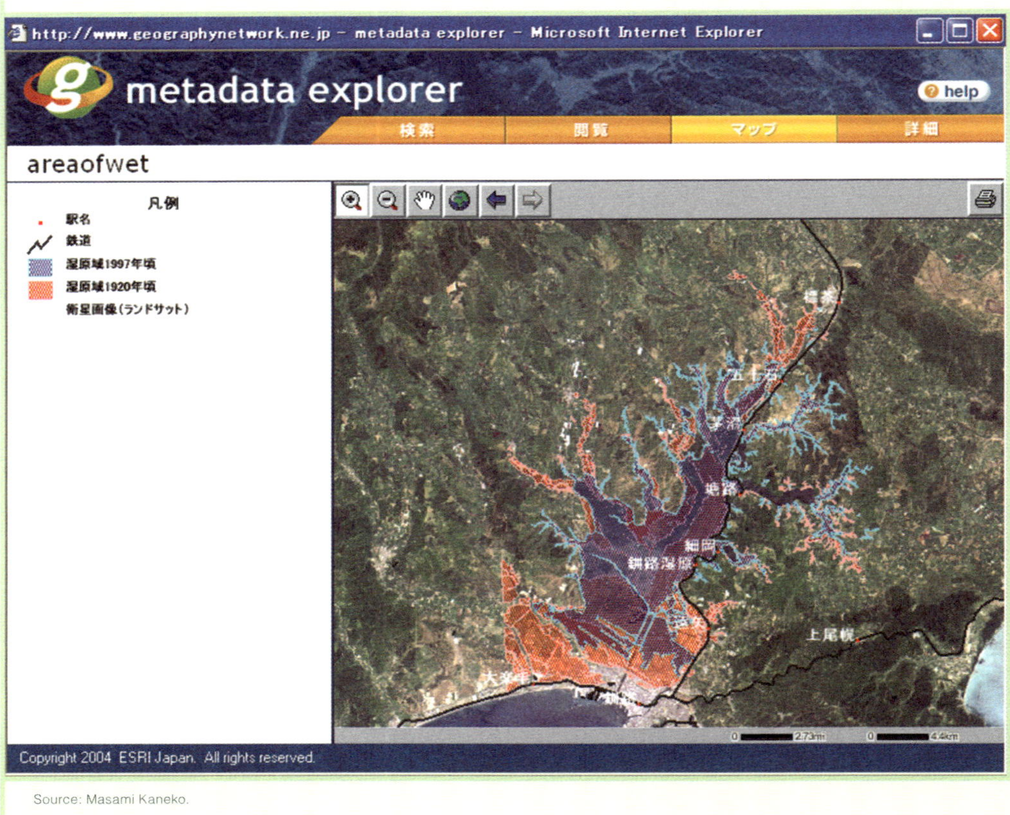

Figure 4.13 This Web site is used by scientists for research work and is also a medium for communicating with the public.

Source: Masami Kaneko.

References

InfoRochester.com. Gardening Resources for Rochester, New York. www.inforochester.com/gardening.htm

International Joint Commission. Rochester Embayment Stage 2, Remedial Action Plan Review, February 12, 2002. www.ijc.org/rel/boards/annex2/reviews/raprefeb12.html

Korfmacher, Karl. 2004. Assessing wetland changes in the Rochester Embayment Area of Concern. Presented at the ESRI User Conference, San Diego, Calif. gis.esri.com/library/userconf/proc04/abstracts/a1264.html

National Audubon Society. How many wetlands have already been destroyed? And why? www.audubon.org/campaign/wetland/destroy.html

Nature Restoration Project in Kushiro Shitsugen Wetland. English-language version of Web site. www.kushiro.env.gr.jp/saisei/english/top_e.html

New York State Department of Environmental Conservation. March 9, 1998. Contamination clean up plans for Great Lakes completed. Press release. www.dec.state.ny.us/website/press/pressrel/1998/98x28.html

New York Department of Environmental Conservation and Monroe County Department of Health. 1997. Rochester Embayment remedial action plan executive summary. Rochester, N.Y.

U.S. Environmental Protection Agency. Great Lakes, Rochester Embayment Area of Concern. www.epa.gov/glnpo/aoc/rochester.html

Resource management 5

Reviving Nebraska's historic Platte River

A tributary of the Missouri River, the Platte River flows 310 miles through Nebraska. The Platte played a key role in the western expansion of the United States.

Data provided by 2004 ESRI Data & Maps.

Nebraska's historic Platte River was subject to high spring floods, great loads of sediment, and occasional summer droughts. These conditions caused continuous movement of the braided channels and sandbars, resulting in a very broad, shallow, sandy waterway that lacked vegetation. This type of habitat was ideal for multiple species of wildlife.

During the western expansion of the United States, the Platte River and its valley formed a popular route for fur traders, explorers, military personnel, settlers, and opportunists seeking their fortune in the gold fields of California. The river appears on maps dating to the eighteenth century and is identified by several different names, according to the Nebraska State Historical Society. The river has been called *Panis,* derived from the name of the Native American Pawnee tribe. French fur traders called it *LaPlatte*—"The Flat"—and *Nebraska*, derived from a Native American term meaning "flat water." Lewis and Clark documented the mouth of the Platte on their famous expedition and later the river became part of the Oregon Trail.

The modern Platte River is a very different river system than it was 150 years ago when early settlers and gold miners went west. About 3.5 million people depend on the Platte for drinking water, and the Platte irrigates millions of acres of farmland. As much as 90 percent of the habitat in and along the river has been lost, primarily from the effect of many water storage and diversion projects throughout the Platte River Basin, and the associated land development. Today, dense vegetation and forest occupy from 72 to 91 percent of the pre-1900s channel area.

Figure 5.1 The Platte River opposite Platte City near what is now Cozad, Nebraska, in October 1866. Noted Chicago photographer John Carbutt took this picture as part of a series documenting Union Pacific President Thomas C. Durant's excursion to demonstrate the railroad's westward expansion. Durant took more than two hundred guests to a point located on the one hundredth meridian halfway between Chicago and the Rocky Mountains. The photos also show the Platte River before dams diverted water for agricultural, industrial, and municipal uses.

Source: Union Pacific Historical Collection.

Balancing conservation and commerce

The many demands placed upon the Platte have taken a toll on this unique ecosystem and have significantly reduced the amount of habitat available for four endangered species. They consist of three birds—the whooping crane, piping plover, and interior least tern—and a fish—the pallid sturgeon—which makes its home in the Lower Platte River.

Today, conservation and commerce coexist along the Platte as public policy and the courts have sought to balance the needs of competing interests. GIS plays a key role in managing the river's natural resources.

In July 1997, Nebraska, Wyoming, Colorado, and the U.S. Department of the Interior signed an agreement to address several Endangered Species Act (ESA) issues affecting water resource development in the Platte River Basin. The agreement created the Platte River Endangered Species Partnership, which has two main objectives. The first is developing the Platte River Recovery Implementation Program (PRRIP) to maintain, improve, and conserve habitat for the four endangered species that depend on the river. The second objective is to allow existing and new water uses in the Platte River Basin to proceed without additional ESA requirements for the program's four target species.

Source: Platte River EIS Office, Bureau of Reclamation, in conjunction with the Fish and Wildlife Service.

Figure 5.2 The many demands placed upon the modern Platte River have reduced the amount of habitat by about 90 percent for four endangered species. From top to bottom, these are the whooping crane, the piping plover, the interior least tern, and the pallid sturgeon. The animals live in and along the lower portion of the river.

Establishing a baseline

The PRRIP staff set out to establish an ecological baseline for a number of different parameters. In 1998, the U.S. Bureau of Reclamation's Remote Sensing and Geographic Information Group contracted to create a GIS database of land cover and land use along a 90-mile segment of the central Platte River from Lexington, Nebraska, downstream to Chapman, Nebraska. This large GIS database was created for an environmental impact statement (EIS) for the PRRIP.

"Very early on we realized that we needed to have some comprehensive way to characterize the habitat area that is the focus of the program," said Curt Brown, who manages the Platte River EIS Office for the Bureau of Reclamation, Great Plains Region. "We saw from the beginning that we would need a GIS system to help establish that baseline and also to serve as a way to archive a lot of baseline and trend information over time. I think in any setting where your primary focus is management of land, water, or other geographic resources, a GIS system just very naturally becomes the most compelling way to archive data because so much of the data is geographically referenced."

Figure 5.3 GIS plays a significant role in balancing the needs of conservation and commerce along the Platte River.

Photo by Robin Stover, www.nebraskaphotos.net.

Figure 5.4 A detailed 1998 land-cover and land-use map created by the Bureau of Reclamation. Surveys of the river are divided into bridge segments, the areas between the bridges that span the Platte River.

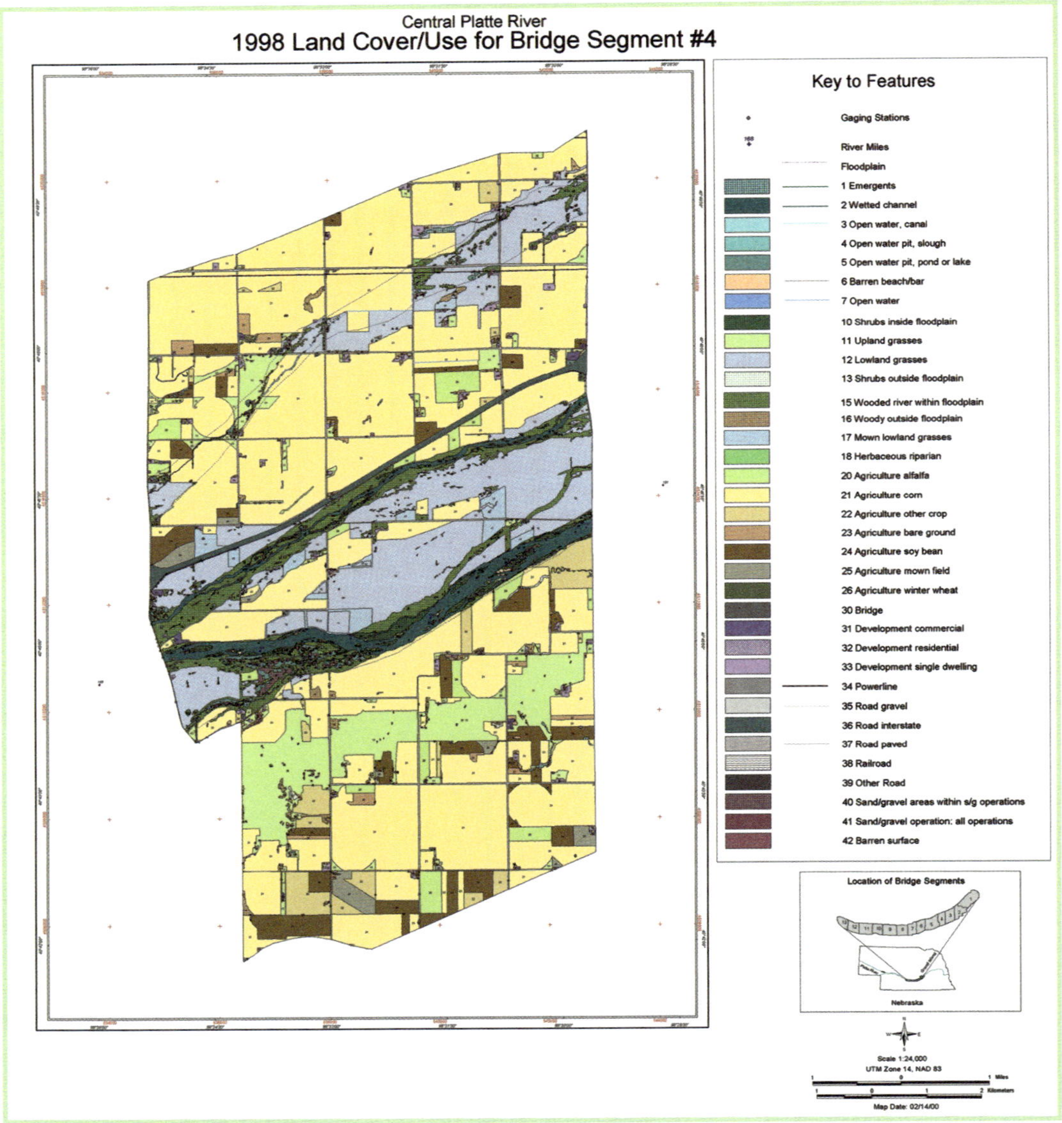

Source: Platte River EIS Office, Bureau of Reclamation, in conjunction with the Fish and Wildlife Service.

Investing in a quality GIS

GIS is the primary way the Platte River EIS office organizes its vast amount of hydrologic, geomorphic, vegetation, and species habitat information. The GIS manages the data for the benefit of the area's four endangered species. The project began in 1998 using ARC/INFO software to create the GIS database and ArcView software to manipulate data and make maps. Information soon began pouring in from various sources, primarily government agencies. Basin surveys in 1982 and 1995 produced good data, but both lacked the detail needed for the environmental impact study baseline.

The GIS project began in October 1998 with initial field inventory by Bureau of Reclamation biologists and GIS specialists. They examined land-cover and land-use elements and assessed the quality of color-infrared (CIR) photography taken in August 1998 for this project. A basemap image was prepared by registering the CIR photos. Used for photointerpretation, the orthophotos were produced at a scale of 1:12,000. To classify plant life, the biologists and specialists followed guidelines set forth in the Federal Geographic Data Committee's vegetation classification standards.

Each orthophoto was covered with Mylar drafting film and systematically interpreted and delineated. Each polygon was labeled with the appropriate map unit number.

The data was converted from raster to vector format after the Mylar sheets were scanned with large-format scanners. ArcEdit software cleaned up line work and assigned attribute values for each polygon. In all, twelve natural vegetation classifications, seven agricultural land-cover types, five surface hydrology categories, fourteen land-use classifications, and a floodplain boundary were interpreted from the 1998 CIR photos.

Base layers of natural features such as river lines, hydrography, and elevation supplemented the 1998 land-cover and land-use database. Also shown were political boundaries such as state borders, county lines, and major cities. The Nebraska Department of Natural Resources and TIGER (Topologically Integrated Geographic Encoding and Referencing) data provided many of the base layers.

"What was not readily available was created by our GIS staff, like land-ownership boundaries that we digitized from maps provided to us by individuals at the Audubon Society, the Nature Conservancy, and water districts," said Joyce Lewallen, GIS Specialist for the Platte River EIS office of the Bureau of Reclamation. "We also worked with hydrologists, biologists, and engineers who would give us coordinates from ground surveys for geographic data like wells, sand dams, and sediment measuring transect. All of the final data products really paid off." The land-cover and land-use database has been very beneficial in the analysis for the EIS, she said.

In addition to the land-cover and land-use database, the Platte River GIS also draws upon an extensive database of transects surveyed along the river. The transects—surveys of water conditions dating from 1989 to the present—cover the 150-mile area between Grand Island and North Platte, Nebraska. GIS was a real benefit for managing, analyzing, and updating this dynamic dataset.

"The transect data were all on paper, and we just basically took everything, including ground photos, sediment surveys, and transect descriptions and put it in a Microsoft Access database so it's easy to query," Lewallen said. "Then, using ArcGIS [software] we mapped the x- and y-coordinates for each transect headpin and connected the headpins with a line feature that represents the transect."

Curt Brown, the Platte River EIS manager, found GIS was perfect for archiving the many different databases and data layers used in the numerous models being run for the EIS. "GIS becomes sort of a nexus between all of the different models," he said.

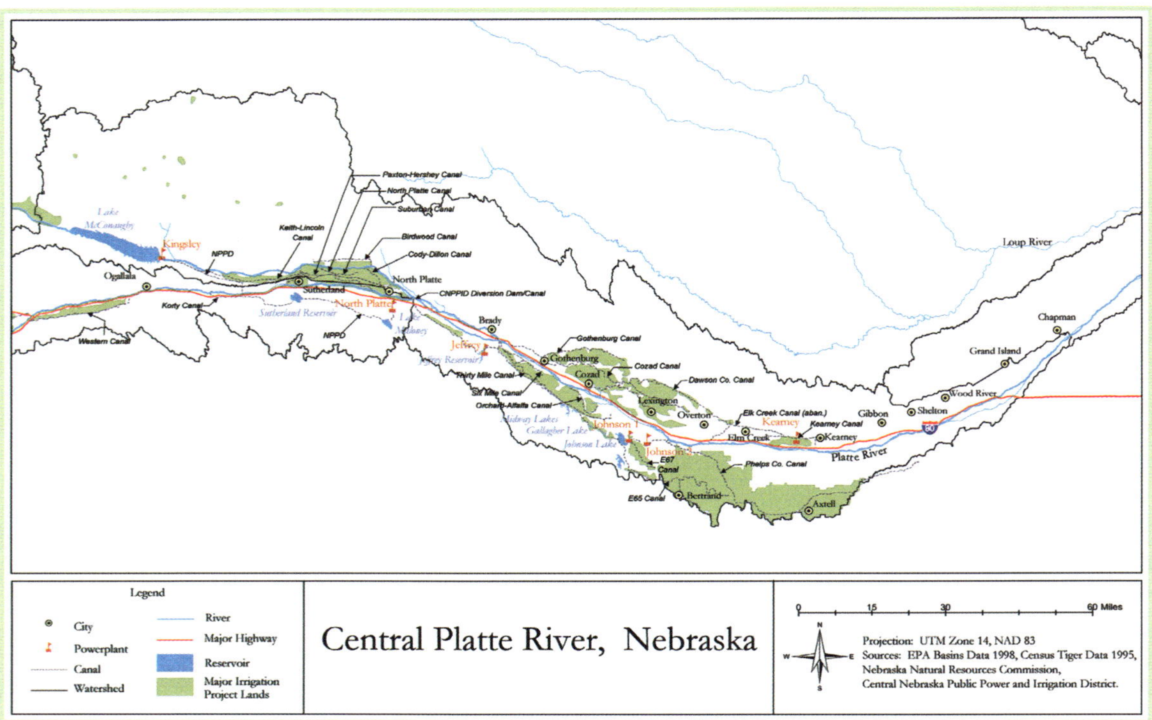

Source: Platte River EIS Office, Bureau of Reclamation, in conjunction with the Fish and Wildlife Service.

Figure 5.5 This map details the major features that define the Central Platte River in Nebraska, including cities, interstate highways, reservoirs, canals, irrigation projects, and power plants.

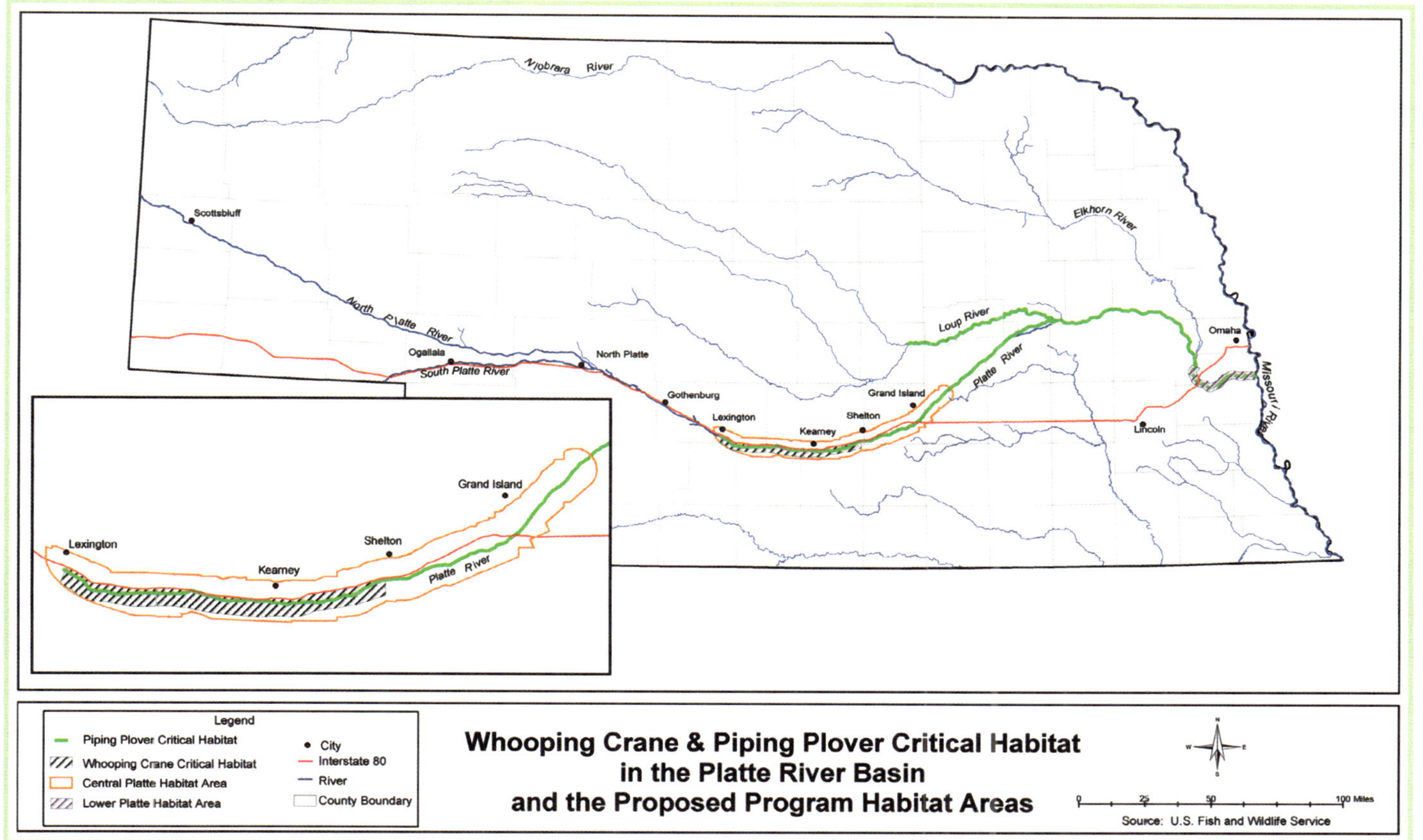

Source: Platte River EIS Office, Bureau of Reclamation, in conjunction with the Fish and Wildlife Service.

Figure 5.6 GIS helped identify habitat critical to the survival of the whooping cranes and piping plover.

Source: Platte River EIS Office, Bureau of Reclamation, in conjunction with the Fish and Wildlife Service.

Figure 5.7 This comparison of aerial photography from 1938 and 1998 of a portion of the Platte River near Lexington, Nebraska, shows the dramatic changes that have taken place over sixty years.

Changing roles

After the land-cover and land-use GIS database was completed in 2000, the GIS shifted to more of a support role, conducting analyses and creating maps for the EIS. The land-cover and land-use database serves as the basis for a majority of the GIS analysis. "Our analyses for the EIS consist of working with a variety of scientists ranging from botanists to river geomorphologists to map various resource management alternatives that will become part of the final environmental impact statement," Lewallen said.

These diverse scientific disciplines frequently intersect, generating new data that is incorporated into the GIS for other uses. For example, a channel width analysis—initially developed for locating suitable whooping crane habitat by identifying the area of active river channel in various width classes—is used in other models to analyze river morphology. A constant demand for new analysis exists, and the depth of information within the system allows for very detailed studies, according to Lewallen.

Currently, the Platte River EIS office is looking into historical trends from 1938 to 1998 to accurately calculate the loss of suitable habitat for the piping plover and least tern. This project uses ArcMap software to calculate the amount of potential habitat from black-and-white aerial photography taken in 1938 and comparing these values to the 1998 land-cover and land-use maps.

Historical maps and aerial images also enable the GIS to perform accurate long-term change detection. One of the data sources acquired by the project included U.S. Geological Survey quad maps from the early 1900s. By digitizing the riverbank boundary lines from the historical maps, the data can be compared with modern maps of the river channel.

In addition to these larger efforts, the Platte River GIS consists of some smaller analyses.

"Using ArcMap [software] we have mapped locations for potential hunting and viewing wildlife blinds, mapped location of built-up sand dams, and located areas of river channel that have islands that have been cleared of vegetation," Lewallen said. As the PRRIP moves into the future, its land-cover and land-use database will be used for individual analyses requiring specific information for developing management alternatives or evaluating adaptive management efforts.

Future of the Platte River GIS

Lewallen found the hours and money invested into creating this land-cover and land-use database to be worthwhile as it has been a foundation for other GIS databases, analyses, and maps.

In addition to making the many studies completed by the EIS possible, the Platte River GIS also helps educate the public. "A majority of the output from our analyses is presented to the public using maps created in ArcMap. These maps help the public understand what it is that we are doing," Lewallen said.

A team with the Bureau of Reclamation is building an intranet site for the Platte River GIS so that other departments within the bureau can make use of the findings, analysis, and data. The Platte River GIS will continue to grow once a recovery plan for the river and its endangered species is implemented. "After a program is adopted, our main focus will be long-term monitoring," Lewallen said.

Figure 5.8 Clearing vegetation from the river is helping to restore parts of the Platte River.

Source: Platte River EIS Office, Bureau of Reclamation in conjunction with the Fish and Wildlife Service.

References

Central Platte River. 1998. Land cover/use mapping project Nebraska.

Committee on Endangered and Threatened Species in the Platte River Basin. 2004. Endangered and threatened species of the Platte River. Washington, D.C.: The National Academies Press.

Nebraska State Historical Society. Platte River history. www.nebraskahistory.org/publish/markers/texts/platte_river_history.htm

Platte River Endangered Species Partnership. www.platteriver.org

Platte River Whooping Crane Maintenance Trust, Inc. www.whoopingcrane.org/index2.html

U.S. Department of the Interior. 2000. Bureau of Reclamation, Fish, and Wildlife Service. Technical report of the Platte River EIS team.

Brownfields 6

Reclaiming land at England's epicenter of the Industrial Revolution

Stockton-on-Tees' history stretches back to Anglo-Saxon settlements established around AD 600 and construction of a castle in the twelfth century by the Prince Bishops of Durham. The area was one of the epicenters of the Industrial Revolution.

Data provided by 2004 ESRI Data & Maps.

The Industrial Revolution arrived early in historic Stockton-on-Tees, a major seaport on England's northeastern coast. Since the seventeenth century, ships sailing from Stockton-on-Tees carried raw materials and agricultural goods produced in northern England to ports around the globe. In 1825, the world's first public railroad, the Stockton and Darlington Railway, began operation in the area. Railways and industry urbanized the agricultural area as trains hauled coal and lead from Britain's interior to the coast. The heavy industries that followed the railways into the area attracted workers from the countryside and the population grew. During the past two centuries, Stockton-on-Tees has at various times been a center of salt making, mining, brick and tile manufacturing, iron and steel fabrication, and shipbuilding, as well as engineering and chemical production.

Today, the Stockton-on-Tees Borough Council (SoTBC) governs a mixture of busy town centers, urban residential areas, and picturesque villages spread across 20,400 hectares (79 square miles) with a population of approximately 180,000. However, two hundred years of industrialization have left behind a legacy of contaminated industrial sites and derelict real estate.

Figure 6.1 Railways, such as this in the community of Yarm within the Stockton-on-Tees Borough Council, played a key role in the industrial development of the area. The world's first public railway began operating here in 1825.

Photo by Peter Mernagh, Stockton-on-Tees Borough Council.

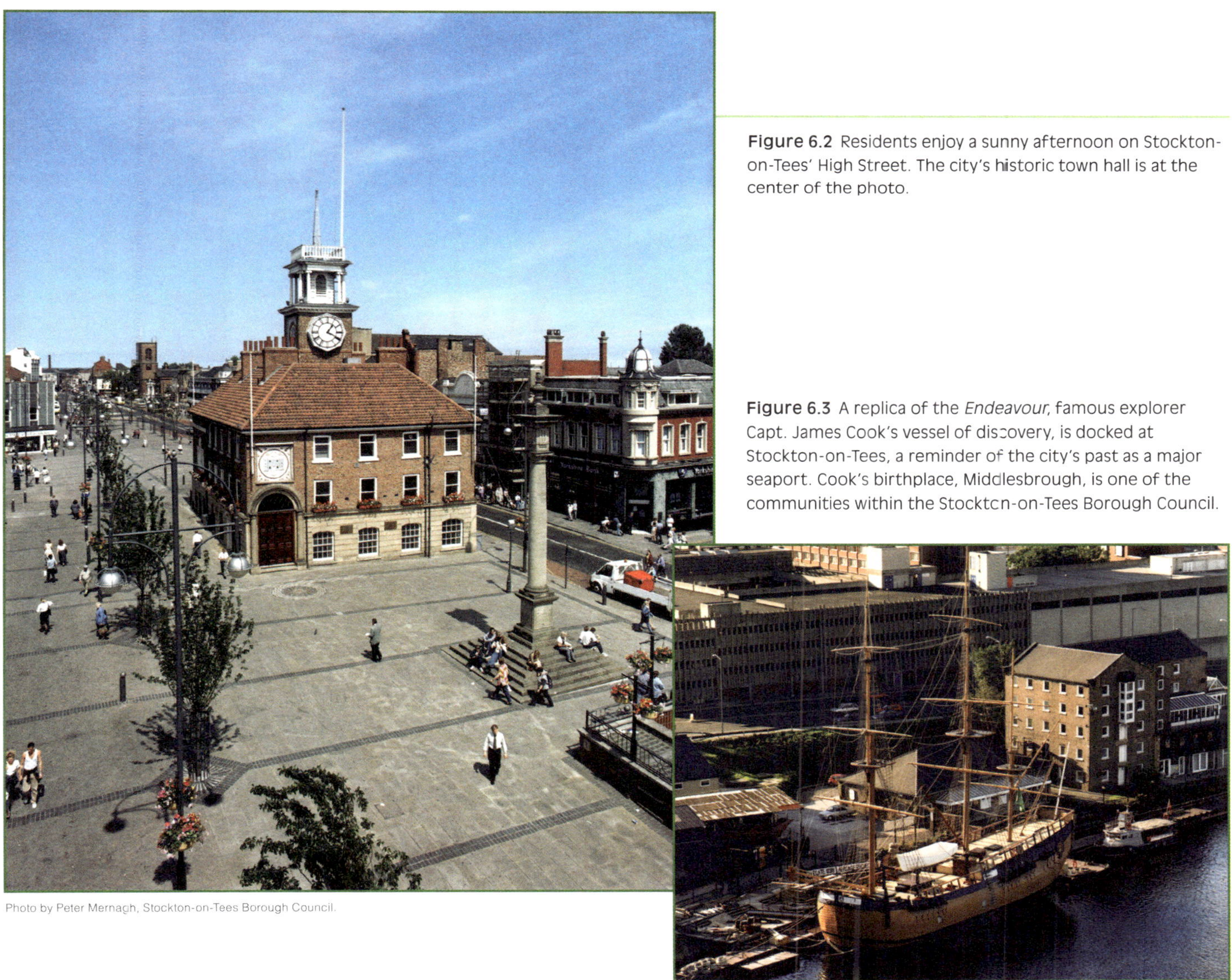

Figure 6.2 Residents enjoy a sunny afternoon on Stockton-on-Tees' High Street. The city's historic town hall is at the center of the photo.

Photo by Peter Mernagh, Stockton-on-Tees Borough Council.

Figure 6.3 A replica of the *Endeavour*, famous explorer Capt. James Cook's vessel of discovery, is docked at Stockton-on-Tees, a reminder of the city's past as a major seaport. Cook's birthplace, Middlesbrough, is one of the communities within the Stockton-on-Tees Borough Council.

Photo by Peter Mernagh, Stockton-on-Tees Borough Council.

Photo by Peter Mernagh, Stockton-on-Tees Borough Council.

Figure 6.4 This aerial view of the Tees River corridor through Stockton-on-Tees shows the mixture of industrial, commercial, and residential land uses in the area. Transportation still plays a key role with a major rail center and motorway visible in the upper right of the photo.

GIS as a time machine

GIS is helping local officials revitalize this historic seaport by tracking sites that may have contamination problems, clearing the way for redevelopment. A government policy in the United Kingdom, known as Section 57, requires that 60 percent of all new urban development take place on brownfields—previously developed—sites. The legislation, which applies to all local governments in the United Kingdom, mandates strategic reviews designed to identify sites potentially contaminated with pollutants and toxins. The law requires local authorities to conduct inspections, assess risks, and identify and remediate contaminated sites. The law also requires a "traceable, rational, ordered, and efficient" process for assessing all land in a consistent and repeatable manner.

Identifying and surveying all potentially contaminated sites within the SoTBC area was a complex undertaking that required as much focus on the past as the present. Thanks to a wealth of historical data, the SoTBC's contaminated lands GIS functions as a time machine that can clearly show decades of changes. Facing a tough mandate set forth by the national legislation, the Stockton-on-Tees Borough Council chose GIS as the tool for dealing with this complex process. The Jacobs Babtie Group of Glasgow, Scotland, was tapped to work with the borough's existing ArcView system to create a modified version of its contaminated land mapper, which is known as Section 57 CoreTools.

Section 57 CoreTools uses established risk assessment principles and the spatial analytical capabilities of GIS to automatically assess the whole Stockton-on-Tees area, according to Stuart Gillies, former GIS manager for Jacobs Babtie on the Stockton-on-Tees project. This system ranks potentially contaminated sites by risk, thus allowing the SoTBC to decide what areas require in-depth inspections, explained Steve Smith, principal environmental health officer at the Stockton-on-Tees Borough.

But creating the system was a complex job. Many potentially contaminated sites existed, and large amounts of detailed data associated with the sites required processing. More than 120 datasets used within Section 57 CoreTools included detailed maps of the area dating to the 1850s created by Britain's Ordnance Survey (OS). Aerial photos dated as far back as the 1920s. Using these historical basemaps was critical for an area like Stockton, where the history of industrial pollution goes back nearly two centuries.

The OS conducted several detailed surveys of all of Great Britain between the 1850s and the present. These historical maps, combined with the other datasets, allow the model to become a sort of geographic time machine. It can perform change detection over a wide area and a long period of time.

"You can actually see 150 years of history in these urban areas," Gillies said.

Building an efficient model

Gillies started on the project in early 2001, and he quickly realized the situation was perfect for creating a GIS model that would automate the land classification process. He created the model while the SoTBC was manually performing the same function using GIS. "It took someone nearly six weeks to manually go through all of the datasets, look at the various layers, and do all of the calculations needed for assigning the weighted values," Gillies said. "When we did the same thing by running the model it took four minutes."

The first component of Section 57 CoreTools uses ArcView software tools for data management, modeling, and mapping and the geodatabase functions of ArcInfo software to run analyses of a critical data source. The second is the seamlessly linked SoTBC Land Quality Management Database (LQMD) that records information related to each potentially contaminated site. The third critical component of Section 57 CoreTools is a comprehensive metadatabase.

"We have a metadatabase that is the foundation for all of the work we do [in GIS]," Gillies said. "Rather than the metadata being associated in a file beside the data itself, it's centralized and our users don't need to worry about where the data resides on the server, because that's all dealt with in the metadatabase and the model runs on top of it." Section 57 CoreTools uses more than 120 digital datasets split into three categories: sources, pathways, and receptors. Data management was critical to creating a workable GIS. "The model has a spatial database engine behind it,

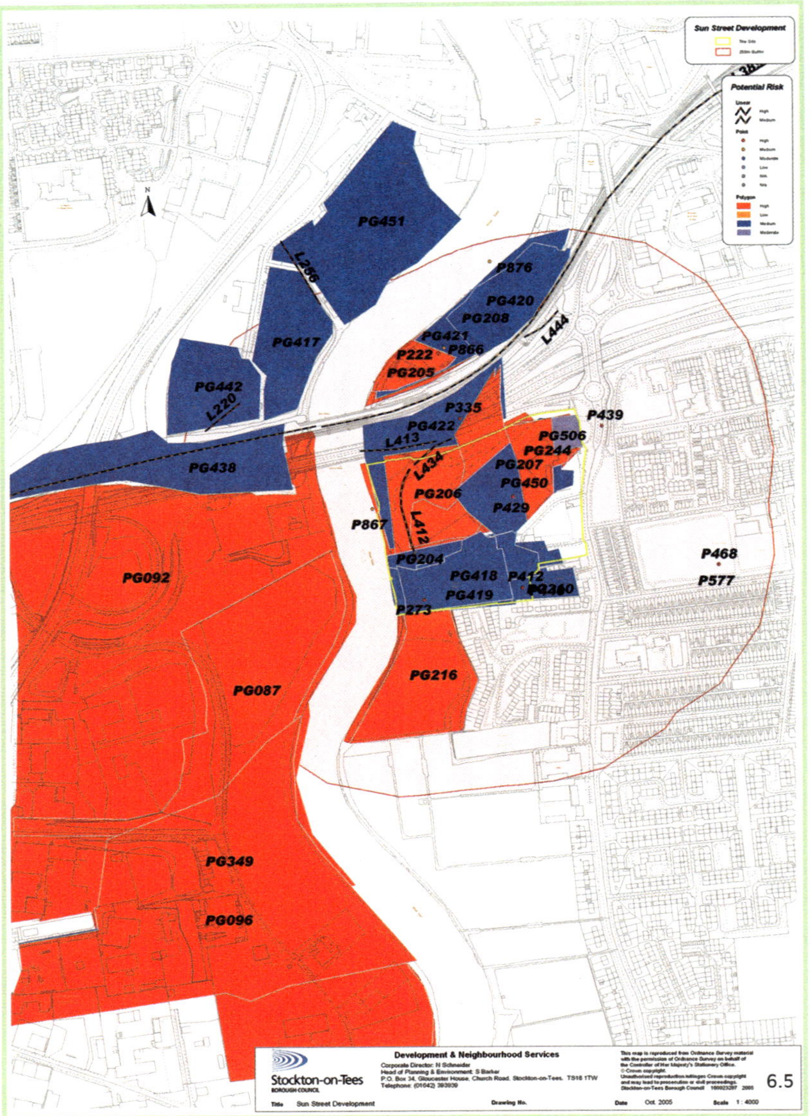

and it does proximity analysis and things like that," Gillies said.

The source category includes areas with historical land uses that may have caused contamination, including heavy industries such as shipbuilding, chemical production, mining, and landfills. Pathways, such as surface water or geological features, allow contaminants to travel between a polluted source site and a receptor. Pathways may include surface water or geological features. Receptors are the destination of the contamination. Receptors are anything that the pollution could harm, including people, wildlife, and environmentally sensitive areas, such as rivers and wetlands. "When you're assessing any site, you're looking for a source of pollution, you look for a viable pathway, and you look for a receptor," Smith said. "If you have a source, but you don't have a receptor, then you don't have a contaminated site."

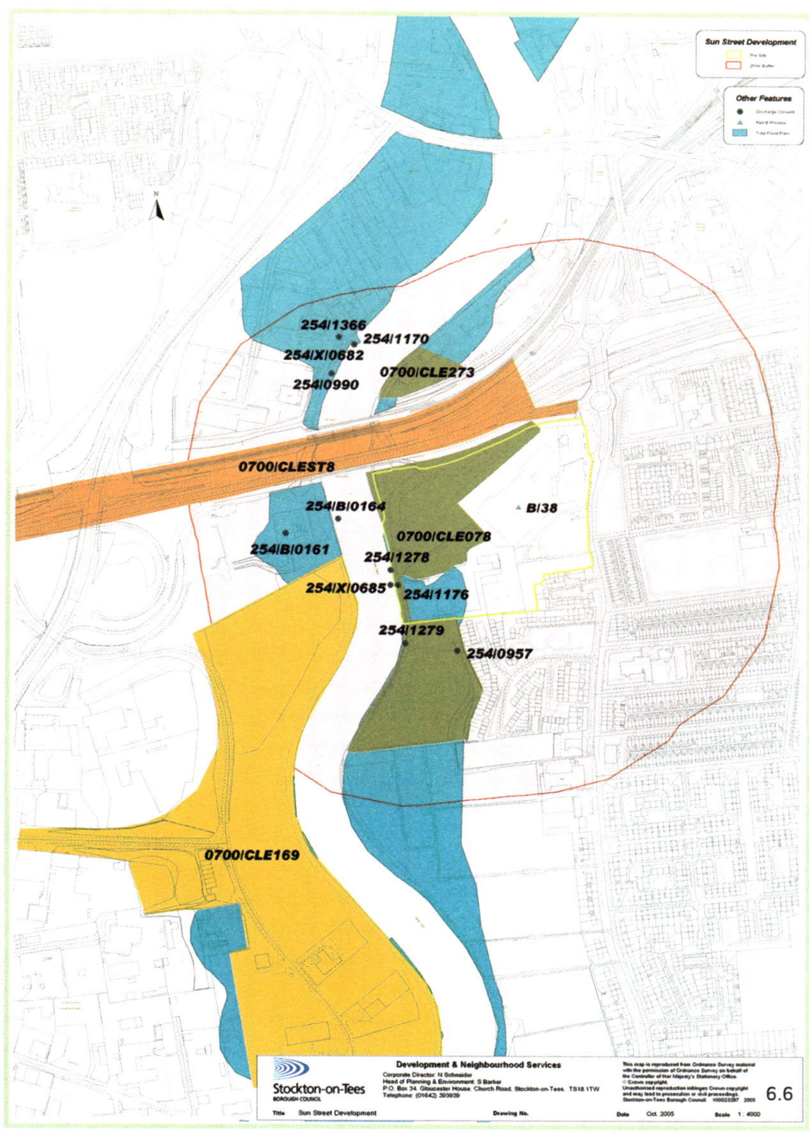

Figures 6.5 and 6.6 Two maps used to analyze potential risks associated with redeveloping a large property in Stockton-on-Tees. The area is being considered for a possible residential development and the risks are from contamination from hydrocarbons and heavy metals. The maps have control numbers and letters that refer to points (P) and lines (L).

Ranking the land

Section 57 CoreTools customizes ArcView software to automate the risk assessment and prioritization process for each piece of land surveyed. The model creates ranked lists of contaminated and potentially contaminated land by applying the spatial analytical capabilities of GIS using certain established source-pathway-receptor risk assessment principles. Potential pollution sources receive weighted scores based upon the potential severity of the site's problems. Pathways are ranked based on the potential for movement of pollutants. Receptors acquire scores based on contamination sensitivity.

The purpose of the resulting rankings is not to assign an actual risk measurement to each site, according to Smith. The rankings measure relative risk from a contamination source to any potential receptors. "There may be chemicals there, but we look at a suitable-for-use scenario," he said. "Just because there's chemicals on the land, that doesn't necessarily mean it's a contaminated site."

Consistently ranking potentially contaminated sites allows the SoTBC to decide where to focus its surveys and remediation efforts. Once each site was ranked, the SoTBC began surveying the sites and building the LQMD using Microsoft Access software. "We had 'round about two thousand sites in the list and they were all given a score," Smith said. The British Department of the Environment divided the two thousand sites into four priority categories following established guidelines. The top priority sites were screened and ground truthing was conducted to ensure that the maps coincided with reality at each site.

Environmental epochs

Smith referred to the development of industrialized Stockton in terms of historical epochs. What appeared on maps as a quarry during one epoch may become a landfill in the next, but an investigation would confirm the status of the land. Within the LQMD, input forms categorized by industrial use record surveys for each potential contaminant that might be present at a particular site. The LQMD contains about twenty-five different pieces of information for each site, including all site investigation reports. Because the LQMD links to shapefiles in the ArcView program, conducting analysis and generating maps directly from the database is easy, according to Smith.

"We can go in and look at the raw data from the maps and follow a sequence through and actually see the features as they change over time," Smith said. "If you have a building, you can outline the current site of that building. Then you can overlay your historical maps for the different epochs and see how [the site] has changed."

The data in the GIS sometimes comes from sources such as old books and photos found in local libraries. Aerial photos are also used. Allowing the public and commercial companies access to the SoTBC's contaminated land GIS database through computers in local libraries and via the council's Web site has also been proposed.

Figure 6.7a One of the sites analyzed by the environmental health unit of the Stockton-on-Tees Borough Council using its contaminated land mapper, built using ArcView software and Section 57 CoreTools. This area started out as a clay pit supplying a local brick and tile manufacturing operation in the St. Ann's Hill area of Stockton-on-Tees. This series of historical maps shows how, in a span of 150 years, the clay pit expanded, was abandoned, became a landfill, and then was turned into a soccer field.

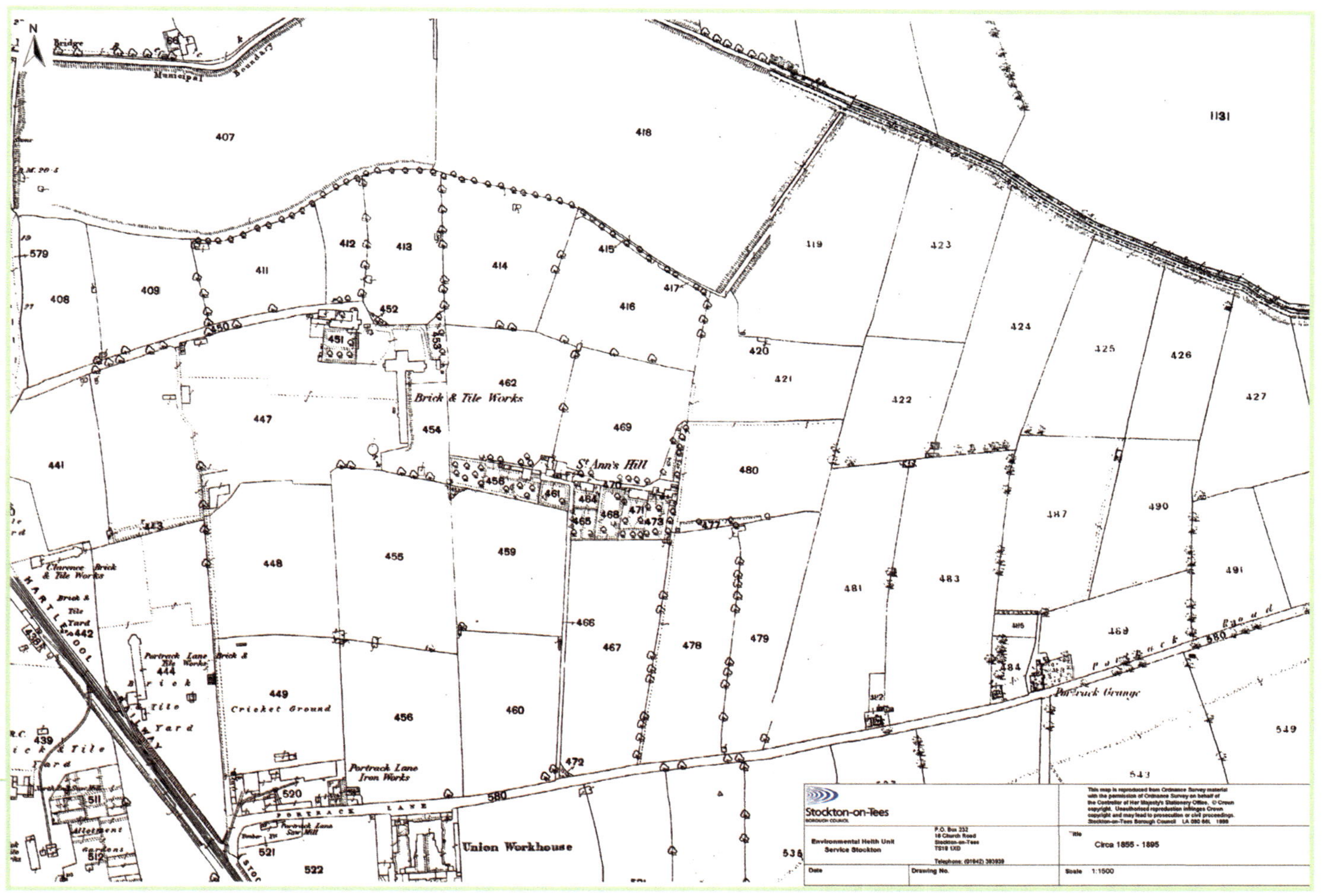
Stockton-on-Tees
Environmental Health Unit Service Stockton
Circa 1855 - 1895
Scale 1:1500
Union Workhouse
Brick & Tile Works
St Ann's Hill
Cricket Ground
Portrack Lane Iron Works
Portrack Grange

Figure 6.7b The brick and tile works and clay pit, circa 1897 to 1899.

Figure 6.7c The brick and tile works and clay pit, circa 1914 to 1920.

Figure 6.7d The brick and tile works and clay pit, circa 1939 to 1941.

Figure 6.7e The clay pit, circa 1950 to 1992. From 1960 until 1971 the clay pit was used as a landfill.

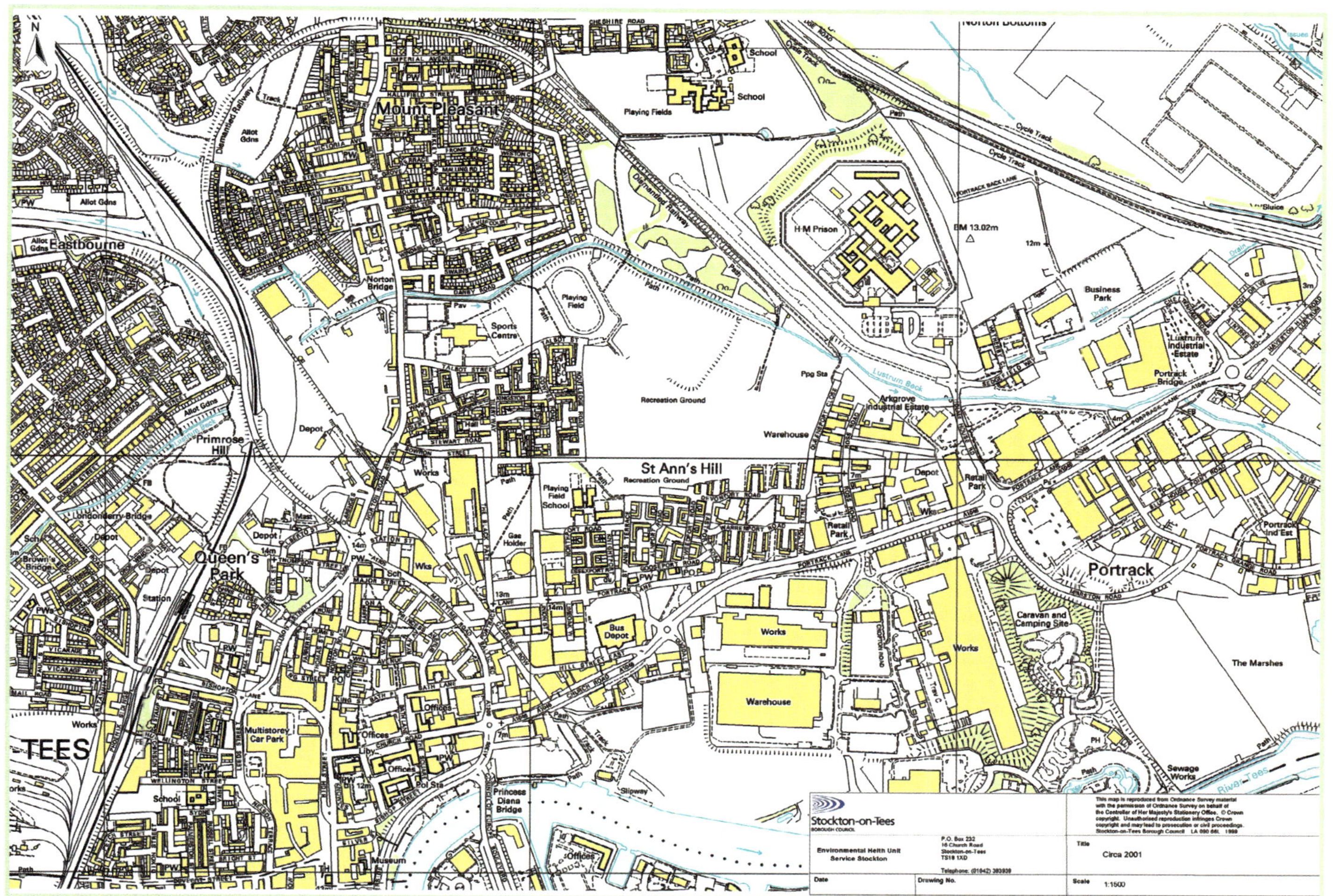

Figure 6.7f The same area, circa 2001, with former clay-pit-turned-landfill converted into a large open recreation area with an adjacent playing field and posts complex.

Turning brownfields into productive property

The SoTBC's work with its contaminated land database is in some ways just beginning. "After the initial inspections, we filled out the database, and we're in the process of reviewing that information to do some further categorization that has to be done manually," Smith said. Some sites are undergoing more in-depth studies to determine the full extent of contamination and what kind of remediation might take place. The SoTBC has much more work to do to determine whether those sites are actually posing a risk of significant harm.

That database continues to grow as more sites are surveyed and problems are resolved. But the future of the GIS is helping transform disused brownfields sites into productive property. A lot of the data gathered for the surveys has other uses. For example, developers are interested in data on individual sites and the SoTBC provides the information to them for a small fee.

There are targets to encourage redevelopment of brownfields areas for housing development. As industrial sites come back into use for housing, the SoTBC must make sure the land is suitable. "Some of these sites, if they stay as they are, may not be contaminated. But once you start putting a housing development on you've got to make sure they don't become contaminated," Smith said. For example, when a new housing development is planned, the SoTBC will check addresses using the GIS to see if there are any contamination issues that need to be dealt with before construction. Planning conditions imposed on developers ensure that remediation work meets required standards and the land is suitable for use. SoTBC values the planning system as an important tool for dealing with potentially contaminated land.

Building the SoTBC's contaminated lands database was a big undertaking. However, managing and remediating polluted land in an area with such a rich and complex history was simplified with GIS. "This is a complex process," Gillies said. "By using GIS, everything has been dealt with in a consistent, scientific way across the whole council area. All of the data used was publicly available, so there's no hidden agenda, and it means the process is open, transparent, auditable, and defensible."

The contaminated lands database helped the SoTBC simplify what could have been a nightmare attempt to meet its obligations to national environmental laws. "You couldn't do it without some sort of system," Smith said. "It was almost a quantum leap from where we were to where we've ended up, and it allowed us to fulfill our legal obligations in an efficient manner."

Figure 6.8 The footprints of modern residences and garages superimposed over a historical map, circa 1914 to 1920, showing the location of a former gas works that supplied power and light for homes, commercial premises, and street lighting. In addition to consulting historical maps, the Stockton-on-Tees Environmental Health Unit uses trade directories that detail possible pollutants for given land uses. This allows environmental health officials to develop a conceptual model for each site that includes all possible pollutants.

References

Shields, Barbara. 2004. Stockton-on-Tees, United Kingdom, identifies and remediates potentially contaminated land with GIS. *ArcNews* (Summer). www.esri.com/news/arcnews/summer04articles/stockton-on-tees.html

Simpson, David. 2001. Northeast England history pages, Stockton-on-Tees. www.thenortheast.fsnet.co.uk/Stockton.htm

Stockton-on-Tees Borough Council. www.stockton.gov.uk

Change detection 7

Monitoring habitat in California's Central Valley

The Central Valley of California is one of the most productive farming regions on earth.

Data provided by 2004 ESRI Data & Maps.

The federal government provides water to more than a million households and most of the farmland in California's Central Valley. Much of the nation's oranges, vegetables, nuts, and cotton are grown in the valley, one of the most fertile farming regions on earth.

Each year, some 7 million acre-feet of water flow to the Central Valley's farms and towns via the federally built Central Valley Project (CVP). The CVP is a system of 500 miles of canals, twenty dams and reservoirs, eleven power plants, and other facilities in California's Sacramento and San Joaquin Valleys. The Bureau of Reclamation, an agency of the U.S. Department of the Interior, operates the CVP and oversees contracts stipulating the quantity and price of water.

However, this large-scale system affects numerous species of flora and fauna as it distributes water from northern California's rivers. In 1992, the U.S. Congress enacted the Central Valley Project Improvement Act to protect and restore fish and wildlife within the CVP. The law requires that 800,000 acre-feet of water annually be used for such conservation purposes.

Since January 2000, the Bureau of Reclamation and U.S. Fish and Wildlife Service have operated the Central Valley Habitat Monitoring (CVHM) program. Its purpose is to map, detect, and monitor land-cover and habitat changes in California's Central Valley caused by the distribution of water from federal water contracts. A Bureau of Reclamation GIS team based in Sacramento leads the mapping program for the change-detection efforts.

Figure 7.1 About a third of the nation's produce, such as grapes and tree fruits, along with significant amounts of cotton are grown in the Central Valley. These crops depend on water provided through contracts with the federal government and delivered by the complex Central Valley Project's canal system.

The CVHM program helps assess the effects of converting agricultural water supplies to municipal and industrial uses. It also provides sufficient water for Pacific Flyway wetlands essential for migratory birds and for fisheries affected by the CVP. One of the program's goals is to identify areas where habitat or land-cover remediation may be necessary to protect endangered or threatened species. The CVHM program turned to GIS to develop a tool to monitor habitat and land-cover change from one time period to another, said Barbara Simpson, geographer and GIS analyst for the Division of Environmental Affairs at the Bureau of Reclamation.

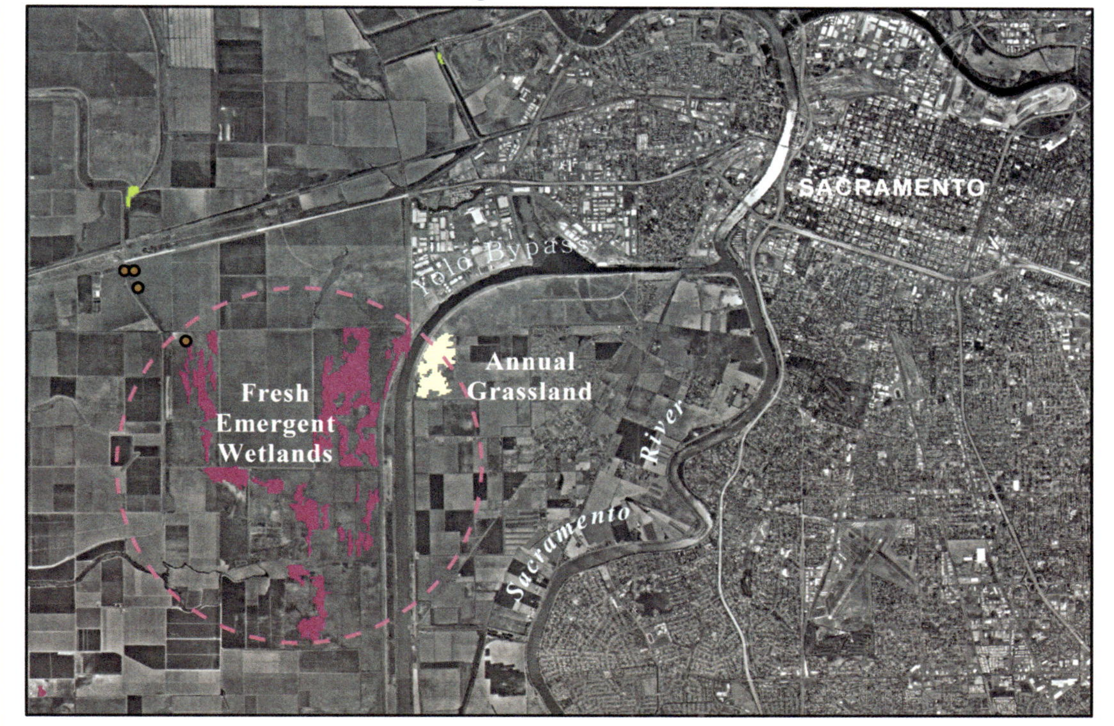

Figure 7.2 This map shows habitat changes in the Yolo Bypass area near Sacramento, California, from 1993 to 2000. It is an example of the Bureau of Reclamation's change-detection projects in the Central Valley.

Source: U.S. Bureau of Reclamation.

Mapping the valley's core

The U.S. Forest Service and the California Department of Forestry have conducted change-detection projects at the edges of the Central Valley. However, the CVHM program was the first change-detection mapping of the valley's core, according to Simpson. One of the factors that set this project apart from some other change-detection efforts was its size. The Central Valley covers the interior heart of the Golden State, extending about 400 miles from the Cascade Mountains near Redding south to the Tehachapi Mountains near Bakersfield. The CVHM program area includes approximately 12.5 million hectares (48,263 square miles) within the valley.

The three-phase program initially developed a land-cover basemap for 1993, when the Central Valley Project

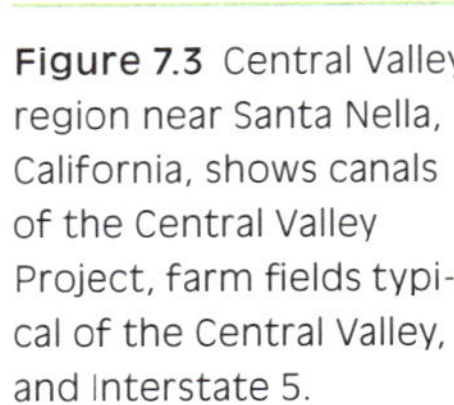

Figure 7.3 Central Valley region near Santa Nella, California, shows canals of the Central Valley Project, farm fields typical of the Central Valley, and Interstate 5.

Source: ESRP, CSU Stanislaus © Endangered Species Recovery Program, California State University, Stanislaus.

Improvement Act took effect. "We started with 1993, because that was the furthest we could go back and still have a good collection of digital data," Simpson said. "We mapped the Central Valley for habitat using satellite imagery, aerial photography, and field work." Existing digital datasets of land cover, land use, and vegetation were used to make the data layers in the 1993 basemap. Developing the map was a major challenge since there was a lot of available data, but none of it precisely overlapped, according to Chris Curlis, senior remote sensing and GIS analyst for the Division of Environmental Affairs at the Bureau of Reclamation.

During phase one, the team devised an initial habitat classification system to serve as a mapping legend. The California Wildlife Habitat Relationships System (WHR), which identifies California's general vegetation groups, served as the basis for the new legend. "As we went through the process, it became apparent that we needed to compile and evaluate all land-cover data available for 1993 and then look at a process for updating that information for our next target year, which was 2000," said David Hansen, soil scientist and GIS specialist for the mid-Pacific Region of the Bureau of Reclamation.

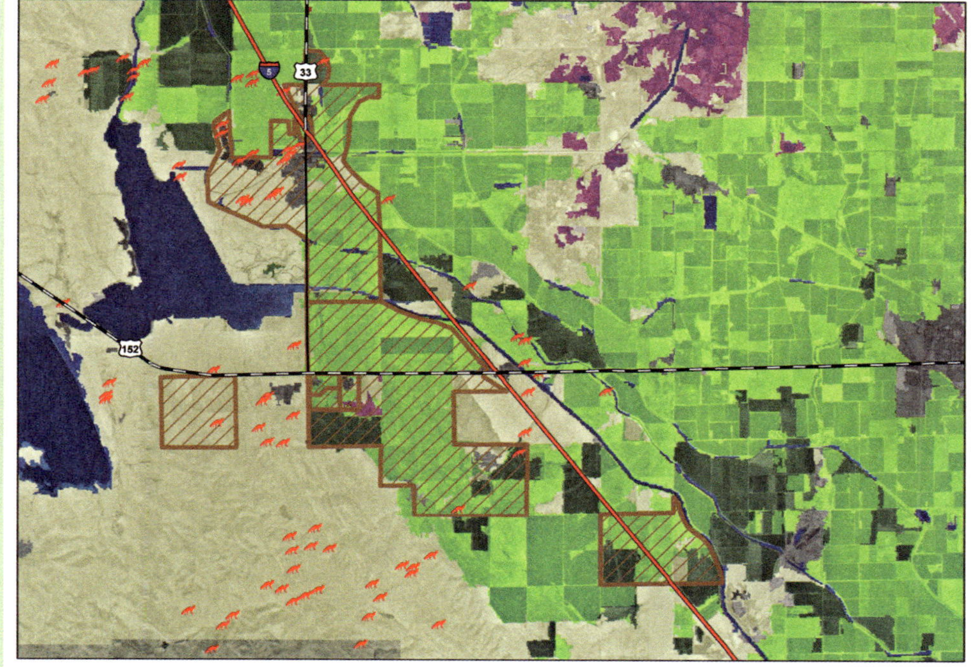

Source: U.S. Bureau of Reclamation.

Figure 7.4 Map of habitat conservation planning for the endangered kit fox in the area around Santa Nella. This is the same area shown from the air in figure 7.3.

A diversity of data sources

"We had six different datasets that were at different resolutions and had different classification schemes," Curlis said. The sources for the primary datasets included

- Wetland mapping by the nonprofit organization Ducks Unlimited
- Land-use mapping by the California Department of Water Resources
- Farmland mapping by the California Department of Conservation
- Data from a native hardwood tree-mapping project led by Norman H. Pillsbury, a professor at California Polytechnic University, San Luis Obispo
- The National Land Cover Data (NLCD) from the U.S. Geological Survey
- Data from the California Gap Analysis project conducted by University of California, Santa Barbara, for the state Department of Fish and Game

Source: U.S. Bureau of Reclamation.

Figure 7.5 Ducks in a wetland in the Yolo Bypass area near Sacramento are among the migratory game birds that use the Central Valley's habitats. Wetlands are rapidly disappearing as development in the valley intensifies.

Ducks Unlimited's mapping project identified wetlands. The Bureau of Reclamation's 1993 basemap used this data to identify water, wetlands, and some riparian habitat types. The Ducks Unlimited wetlands data used satellite images taken during the winter of 1992 to the summer of 1993. Ducks Unlimited's data covered most of the Central Valley floor and became a component in the spectral change analysis. The minimum mapping unit for this data was 1 to 2 hectares (2.5 to 5 acres).

The California Department of Water Resources developed crop and land-use data for 1989 to 1995 from field maps and aerial photographs. The CD-RW mapping, done at 1:24,000 scale, covered only portions of the Central Valley. It identified major crop types and residential, urban, and industrial areas, and showed areas of natural vegetation. However, it did not show specific habitats.

Biannual mapping by the California Department of Conservation (DOC) to identify changes in agricultural land was used in the Bureau of Reclamation's 1993 basemap to identify agricultural and urban areas. Derived from aerial photography at a scale of 1:24,000, the DOC mapping data had a minimum mapping unit of 4 hectares (10 acres) and covered only portions of the project area.

Data mapped by Pillsbury and others was used to identify types of native hardwood trees for the 1993 basemap. This dataset also had its challenges and limitations. The scale of the data varied from 1:24,000 to 1:58,000 and it spanned a long time frame and covered only portions

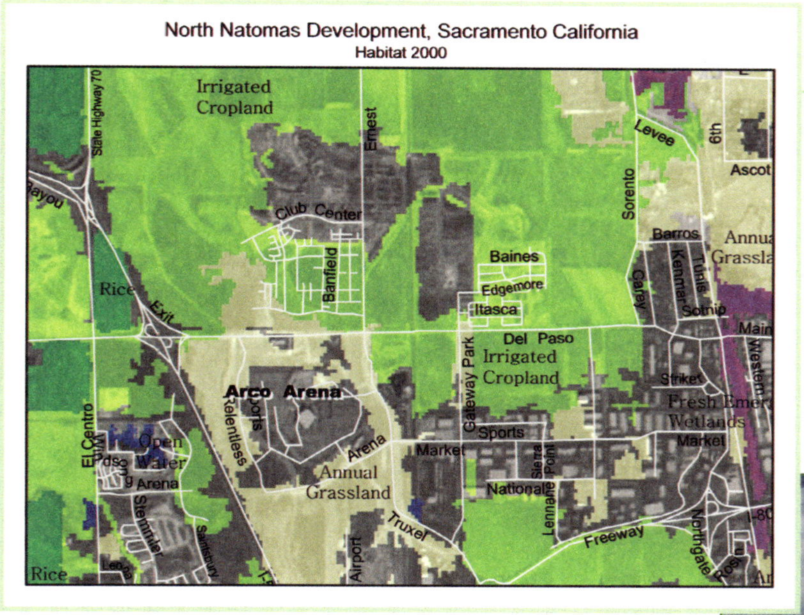

North Natomas Development, Sacramento California
Habitat 2000

Source: U.S. Bureau of Reclamation.

Figure 7.6 This map shows land cover, the extent of development, and wildlife habitat in the North Natomas area of Sacramento for the change-detection study's baseline year of 1993.

of the Central Valley. National Land Cover Data (NLCD), released by the U.S. Geological Survey in 2000 and based on Landsat 5 Thematic Mapper (TM) satellite imagery for 1993, helped provide a more general land-cover classification layer. Each 30-meter pixel of the NLCD carried a label for land cover.

UC Santa Barbara researchers compiled a statewide map of major vegetation or land-cover categories as part of the California Gap Analysis project prepared for the state Department of Fish and Game. It provided the wildlife habitat relationships that became the basis for the 1993 basemap's legend. Although the map had the lowest resolution of the six datasets—a scale of 1:100,000 with a 40-hectare (100-acre) minimum mapping unit—it had a very detailed and comprehensive habitat legend.

Source: U.S. Bureau of Reclamation.

Figure 7.7 A new housing development and landscaping surrounds a wetlands in Sacramento's Natomas area. Rapid development is changing once-rural areas into suburban communities.

Data crosswalks and image segmentation

Combining the six datasets required crosswalks, a process that involves converting classes in each original dataset to one common set of classes. The crosswalking process used ARC Macro Language (AML) for the vector-based polygon data as well as a recording program in ERDAS IMAGINE software for raster-based data such as the NLCD. With the crosswalking process complete, it was possible to assign single labels to the spectral polygons based on each source layer. The end result was a database containing land-cover labels representing the source layer of each spectral polygon.

Figures 7.8, 7.9, and 7.10 These three figures show the sequences involved in producing the images that show changing land uses. Figure 7.8 shows processed light wavelength bands 4, 5, and 3 from 1993 satellite imagery. Figure 7.9 shows the same light wavelength band sequence for satellite imagery from 2000. Figure 7.10 shows the resulting change image with the yellow highlighted area representing changes in land use.

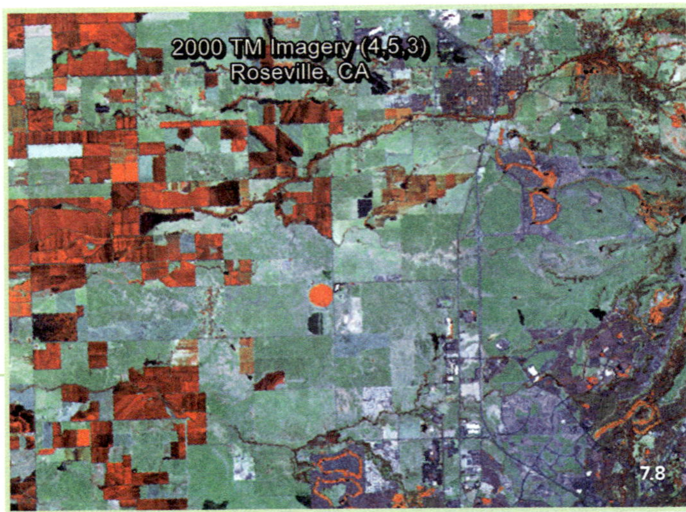

Source: U.S. Bureau of Reclamation.

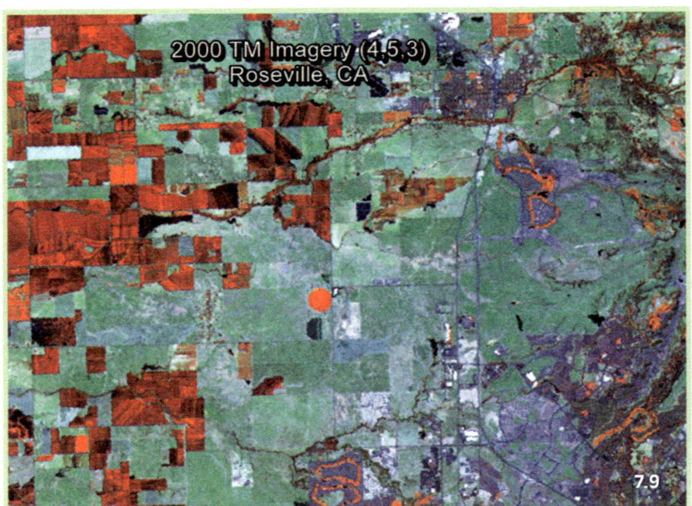

Source: U.S. Bureau of Reclamation.

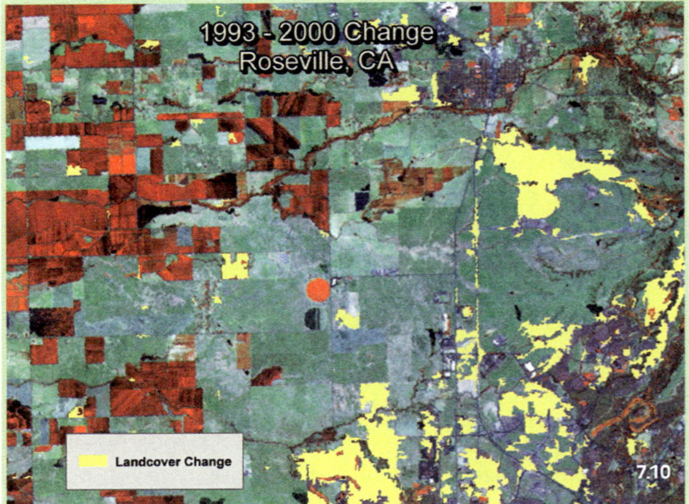

Source: U.S. Bureau of Reclamation.

This allowed for direct comparison of the databases regardless of the original mapping methods.

Merging multiple land-cover datasets for the same geographic area is often difficult because of the differences in classification systems, mapping unit sizes, and methods used for creating polygon boundaries. To overcome this hurdle, the CVHM team used image-segmentation algorithms to create spectral-based polygons. (Image segmentation is an iterative algorithm that aggregates pixels into groups of spectrally similar pixels known as regions. Pixel region boundaries tend to closely resemble actual landscape boundaries, in much the same way as polygons created by interpreting traditional aerial photos.)

Image segmentation provided a way to create polygons from any digital image based only on spectral similarity. The polygons created by the image segmentation method have no labels and instead have unique polygon identifiers. Because of this, it was possible to give land-cover classifications to each polygon based on any data source such as other digital land-cover layers or pixel-level spectral classifications.

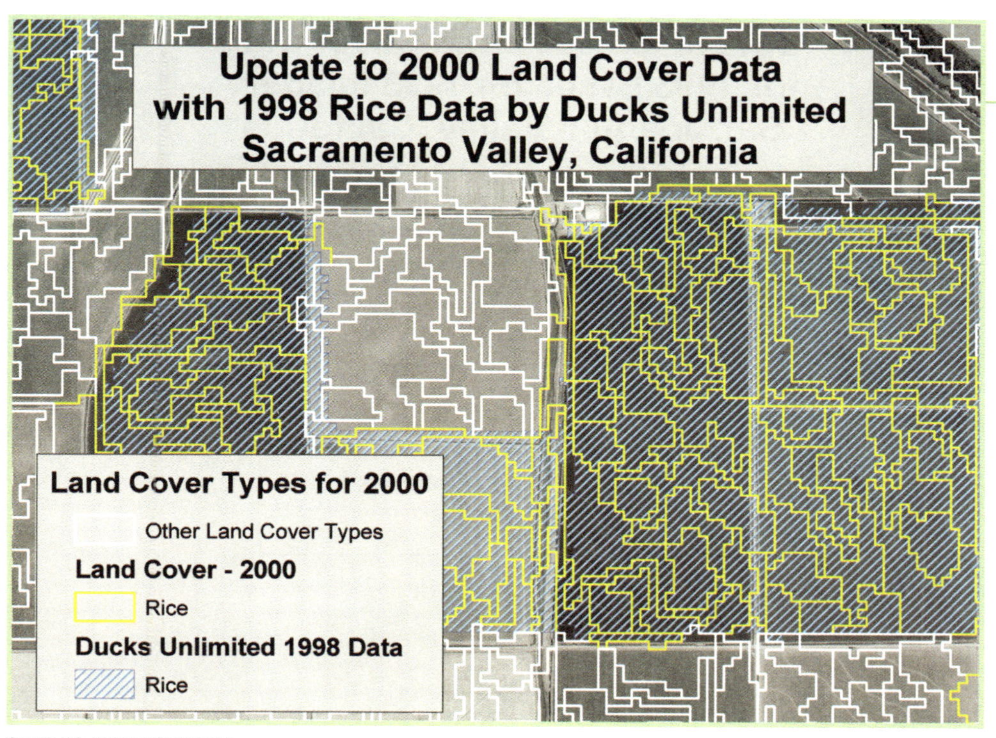

Source: U.S. Bureau of Reclamation.

Figure 7.11 The nonprofit conservation organization Ducks Unlimited mapped rice-producing areas in the Sacramento Valley in 1998. The organization's mapping updated the 2000 change-detection database. The Ducks Unlimited mapping is shown in blue hachures, and spectral polygons are displayed as outlines of the polygons.

Identifying spectral change

Phase two of the CVHM program identified spectral change between the base year of 1993 and 2000. The work started in 2001 as the 1993 basemap neared completion. In this second phase, a pixel-level change map was developed for the 1993 and 2000 images. Landsat TM satellite scenes for the two dates were co-registered and image differencing applied, creating a new image representing spectral change between the 1993 and 2000 Landsat TM imagery. The agricultural mask excluded changes related to farming. The 2000 basemap consists of spectral polygons generated from Landsat TM satellite scenes, with the same techniques used for constructing the 1993 basemap. Image-segmentation algorithms were used to create polygons based on image texture and spectral reflectance.

The 2000 set of spectral polygons was independent of and different from the 1993 spectral polygons. The initial information captured for the 2000 basemap was the habitat mapping labels for 1993 and the percentage of spectral change for each polygon between 1993 and 2000. In addition, several other digital sources for land cover

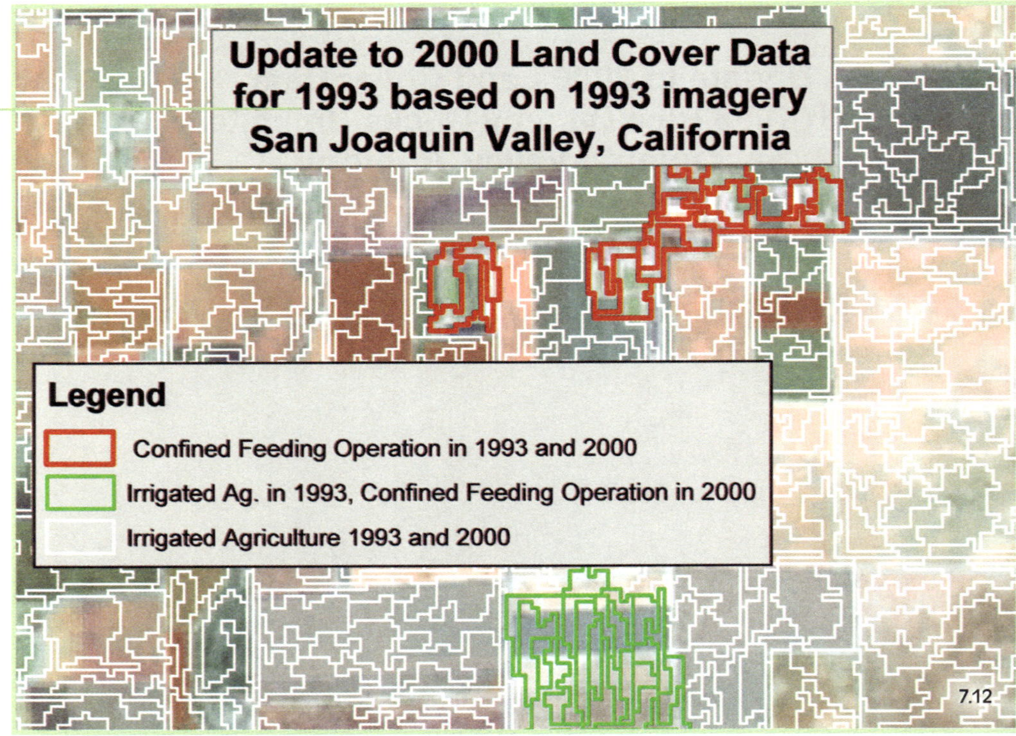

Source: U.S. Bureau of Reclamation.

Figures 7.12 and 7.13 These two maps show updates of the 2000 change-detection database for a portion of the San Joaquin Valley. Figure 7.12 shows confined feeding livestock operations in areas that had previously been identified as irrigated agricultural land. Figure 7.13 shows the same area in 2000 with color imagery in the background. Between 1993 and 2000, the area marked in green had been converted from irrigated agricultural land to a confined feeding operation, a good example of how GIS was used to detect land-use changes.

in the area were available for a comparison between 1993 and 2000. New data incorporated into the 2000 basemap included

- Mapping of land cover by the National Oceanographic and Atmospheric Administration for 1993 based on the Landsat TM images
- Detailed mapping of riparian vegetation conducted by California State University, Chico, in 1998
- Mapping of rice-growing areas in the Sacramento Valley for 1998 conducted by Ducks Unlimited
- Field validation of land-cover mapping conducted by the Endangered Species Recovery Program, Fresno, California, between 1999 and 2001
- Mapping of land cover by the California Department of Forestry and U.S. Forest Service between 1998 and 2001

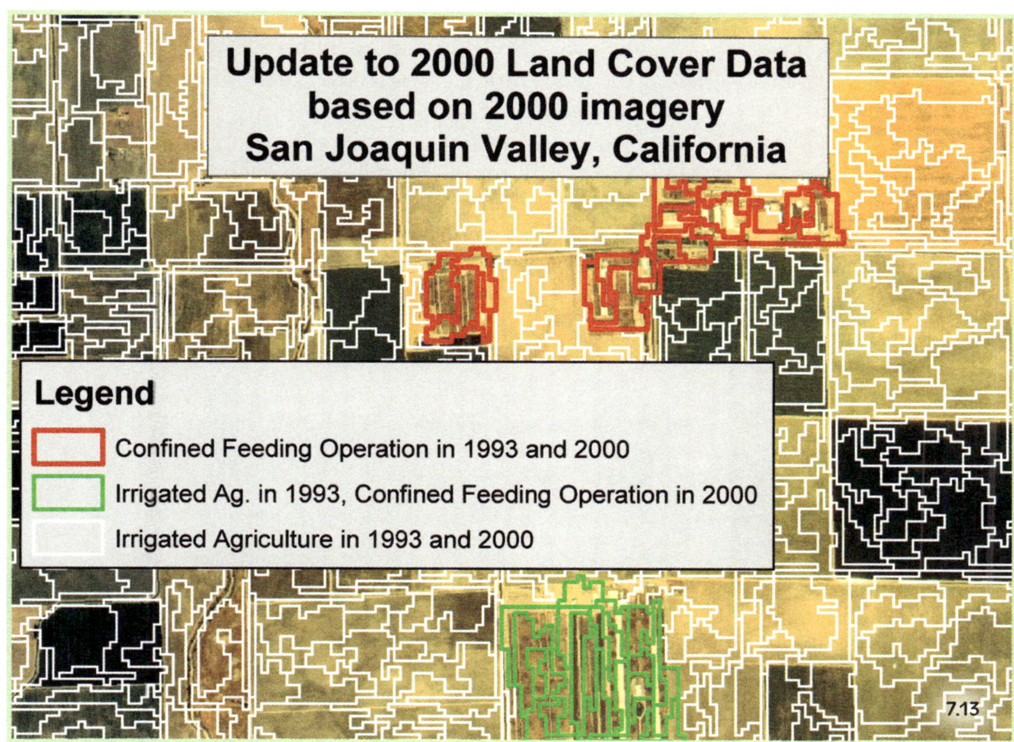

Source: U.S. Bureau of Reclamation.

Preparing for the future

Identifying and evaluating the causes of spectral change along with changes in land cover and habitat from 1993 to 2000 took place during the project's third phase. "We have a lot of reports to generate identifying where change has taken place, within the local water districts where [the Bureau of] Reclamation delivers water in the valley," Simpson said. The GIS team also reevaluated the development of its legend to better reflect changes that have taken place. Changes include where land has been permanently taken out of agriculture, where agricultural land has become fallow, where agricultural land uses have been transferred to municipal and industrial uses, and where land has been purchased for use as wildlife habitat. This phase also involved preparing for the change-detection study for the period between 2000 and 2005.

The CVHM project is a GIS database capable of handling new and updated data as reviews take place.

Source: U.S. Bureau of Reclamation.

As 2005 drew to a close, the Bureau of Reclamation GIS team was conducting accuracy assessments of the 2000 data and preparing for the 2005 change-monitoring review. One of the upgrades planned for the 2005 mapping was the use of Landsat 7 imagery that includes panchromatic data at 15-meter resolution. The new imagery will help sharpen the polygons produced to better depict changes in the landscape.

Software changes are another major issue, according to Simpson. Since it began in 2000, the project has migrated from using ArcInfo and ArcView software to the ArcMap geodatabase format. The mapping effort for 2005 was also to include switching to Definiens Imaging eCognition software to create image segmentation polygons. The eCognition software classifies images based on attributes of the objects in the image rather than the attributes of individual pixels in an image.

The next round of mapping will examine how rapid growth in the Central Valley and shifts in water use have contributed to land-use changes. "Between 2000 and 2005 there's been a lot of change, especially around the urban fringes," Simpson said.

Figure 7.14 This wetland in the Sun River area is an example of the numerous wetlands that dot the Central Valley and provide habitat for migrating birds and other animals.

References

Curlis, Chris, Jeff Milliken, Barbara Simpson, David Hansen. 2001. Putting puzzles together when the pieces don't fit: Data integration. Paper presented at ESRI User Conference, San Diego, Calif., July 9–13.

Hansen, David T., Barbara Simpson, Chris Curlis, Jeff Milliken. 2001. Legend development for a land cover/habitat classification project for the Central Valley of California. Paper presented at ESRI User Conference, San Diego, Calif., July 9–13.

Reclamation: Managing water in the West. 2004. Mid-Pacific Region Year in Review, U.S. Bureau of Reclamation, 55–56.

Simpson, Barbara, Chris Curlis, David T. Hansen, Demetria Adams. 2003. Change in the Central Valley, California, 1993 to 2000. Paper presented at ESRI User Conference, San Diego, Calif., July 7–11.

Urban forest management 8

Seeing the trees for the city in New York

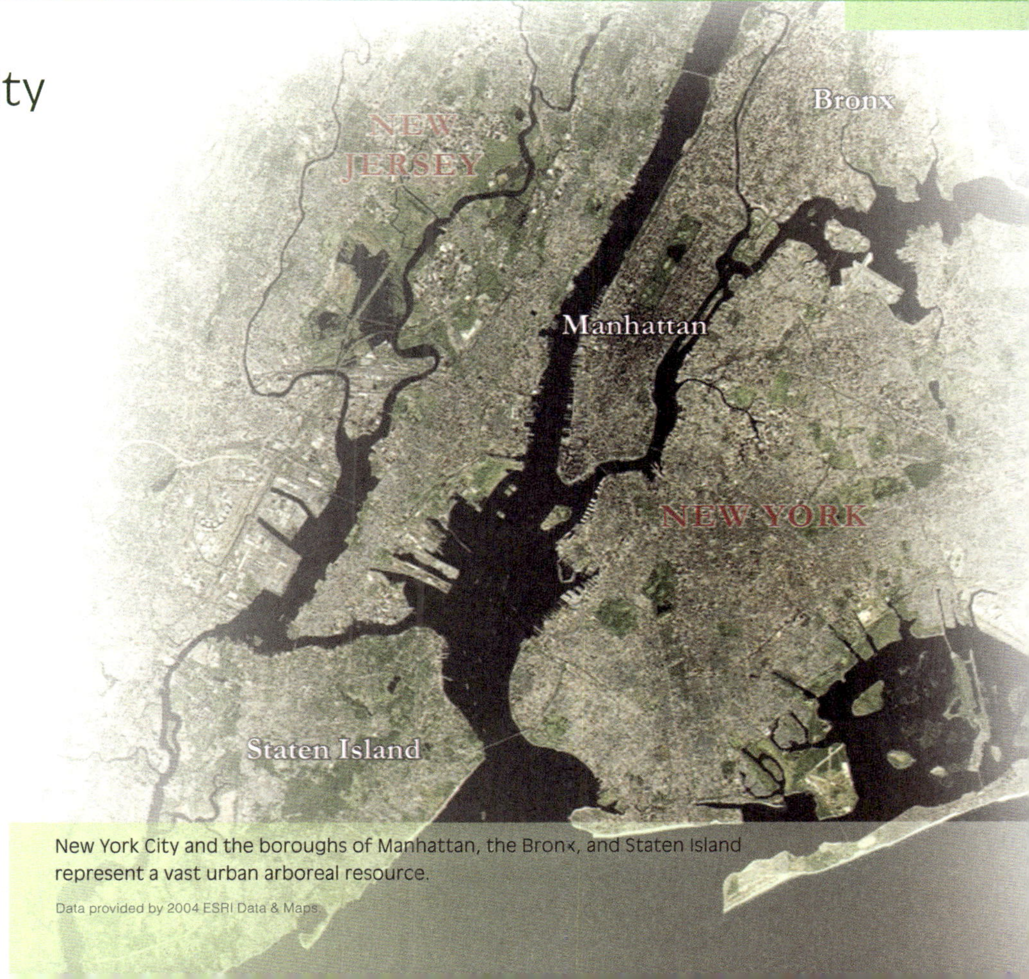

New York City and the boroughs of Manhattan, the Bronx, and Staten Island represent a vast urban arboreal resource.

Data provided by 2004 ESRI Data & Maps.

Step away from New York's Central Park and it can be difficult to see the trees for the city. Yet New York City has an urban forest of more than 5.2 million trees on public and private land. Some 500,000 trees line the Big Apple's bustling streets and contribute to the city's green infrastructure. Until recently, very little data existed documenting this urban arboreal resource.

In early 2001, the U.S. Forest Service took the first step, forming a coalition of more than forty nonprofit organizations, government agencies, businesses, and academic institutions. These groups created the Open Accessible Space Information System (OASIS) cooperative, a Web-based, one-stop, interactive mapping and data application. OASIS provides information about New York City's open spaces with links to other civic infrastructure such as schools, housing, land use, and community planning.

With support from the Forest Service, the groups enlisted the Community Mapping Assistance Project (CMAP) of the New York Public Interest Research Group (NYPIRG) to develop and maintain the OASIS Web site. Powered by ArcIMS software and with substantial support from ESRI and other groups in the coalition, OASIS provides access to more than four-dozen layers of spatial data about the city's environment. OASIS helps people understand their neighborhoods by visualizing spatial data through maps that are accessible via the Internet. It makes information available to communities that cannot afford expensive mapping tools and cannot access the complex pool of government and private sources of open space data.

Figure 8.1 Mention trees and New York City and most people think of Central Park. But New York's five boroughs are home to an urban forest of 5.2 million trees.

Where are New York's trees?

As OASIS creators worked, they found a gap in the many layers of spatial data compiled about New York's five boroughs, according to Lenny Librizzi, assistant director of the Open Spaces Greening Project of the Council on the Environment of New York City. "When we started to look at what types of data that we had, we found one of the things that was missing was a tree layer—specific information about individual trees," Librizzi said. They knew an estimated 500,000 street trees line the thoroughfares of New York's five boroughs but had little scientific data about them. Street trees are vital, but often overlooked, components of urban open spaces and the green infrastructure of cities.

These trees stand out as green islands in a sea of concrete and steel and are most accessible to the public along city streets. The Greening Project wanted to learn how to compile data about street trees. "We wanted to use the information we gathered to come up with a value for these trees and to create data about them that could be displayed on OASIS," said Librizzi.

Figure 8.2 New York City has numerous tree-lined streets. More than 500,000 street trees grow in the Big Apple.

Aided by a $175,000 grant from the Forest Service, plus donations of software and technical assistance from ESRI, a coalition created the Neighborhood Tree Survey Pilot Project as an extension of OASIS. Partners in the project included CMAP, the Council on the Environment of New York City, Trees New York, ESRI, and the State University of New York's College of Environmental Science and Forestry working with the Forest Service's Northeastern Research Station. The project also received in-kind matches from each partner organization and other participants in OASIS such as the New York City Department of Parks and Recreation.

Source: OASISNYC.net.

Figure 8.3 OASIS, the New York City Open Accessible Space Information System cooperative, is an Internet-based, one-stop, interactive mapping and data analysis application designed to enhance the stewardship of open space in New York City. The Web site published the findings of the Neighborhood Tree Survey.

An ambitious agenda

The project had an ambitious agenda, but a limited budget. The Neighborhood Tree Survey Pilot Project's objectives included the following

- Analyzing street tree data collected in the neighborhoods for factors such as carbon storage, air pollution removal, and dollar values
- Providing tree analysis data in a format usable for Web-based OASIS GIS maps
- Developing maps and map layers to select the best locations to plant trees and calculate the benefits and values of street trees
- Developing new urban forest management technologies for New York and communities nationwide

"We were covering a lot of bases, but it was all tree-related," Librizzi said. To reach the goals and stay within budget, the project's geographic scope was limited to three neighborhoods, one each in the south Bronx, the Lower East Side of Manhattan, and New Brighton in Staten Island. The survey's organizers also tapped into New York's legion of Citizen Pruners to help gather data. Citizen Pruners are volunteers specially trained by the nonprofit organization Trees New York to care for street trees. Trees New York's five-session training class bestows special legal status on the Citizen Pruners by granting a "license to prune."

Figure 8.4 An interactive map of the Street Tree Inventory pilot project's study area is displayed on the OASIS Web site.

Source: OASISNYC.net.

Citizen Pruners licensed to trim

"It's actually illegal for people to [prune] the street trees in New York City, even if they planted one in front of their house," Librizzi explained. Trees planted in the strip between the street and the sidewalk, known as the tree pit, are property of the city. Only a licensed arborist or Citizen Pruner is allowed to do any cutting or pruning. "We decided that the people to work with were people that had taken the Citizen Pruner course in the past and already knew something about trees," Librizzi said.

The project organizers recruited Citizen Pruners, training them to measure, categorize, and map street trees. They developed a three-class course to teach the Citizen Pruners how to collect the data. The classes introduced the survey participants to basic GIS concepts and the uses of the data they would collect. "We wanted them to understand that the information they were gathering would actually be used for something and wouldn't just go into some report that would sit on a shelf somewhere," Librizzi said.

One class covered measuring trees, including circumference, trunk diameter, tree height, width of canopy, distance from the ground to the bottom of the canopy, and estimated foliage density. Among the tools the students used were special-diameter tape measures designed for trees and a type of ruler known as a Biltmore stick for measuring tree height.

Photo by Lenny Librizzi, Council on the Environment of New York City.

Figure 8.5 Students and a teacher from PS 45 in Staten Island use Biltmore sticks to measure the height of a street tree. Part of the Street Tree Inventory's mission included an educational outreach component.

Location, location, location

"The other key part was location of the trees," Librizzi said. New York's concrete canyons, where buildings typically tower thirty stories or more, rendered most GPS units ineffective. "Most of the GPS units we could afford had an accuracy of 10 to 30 feet. That's not a problem if you're out in the woods, but in an urban area if you're 30 feet off, that could be across the street from your actual location," Librizzi said.

Because of the shadows cast by tall buildings, most aerial and satellite photos often inaccurately depict the street tree population. Without GPS, accurate tree location was enhanced by the high-resolution basemaps used for the special ArcPad data-collecting application and the printed maps used by the survey participants.

New York City's government has made a substantial investment in developing a high-resolution basemap that includes, among other things, accurate street widths and curb cuts and also accurate building footprints, according to Steve Romalewski, director of NYPIRG's Community Mapping Assistance Project. "We thought that [using maps created from the OASIS basemaps] was more accurate than using GPS units to plot the location of trees," he said.

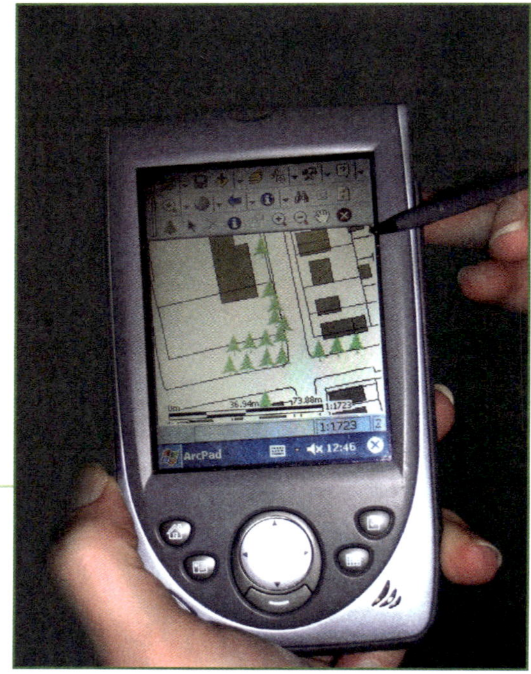

Figure 8.6 Handheld PDAs run a customized ArcPad application used to gather some of the data for the Street Tree Inventory project.

Photo by Lenny Librizzi, Council on the Environment of New York City.

Crunching the numbers

Scientists at the Forest Service's Northeastern Research Station in Syracuse, New York, used the gathered survey data to come up with dollar values for each tree and determine how much carbon and pollution the trees remove from the air. "We had high-resolution digital images flown for all of New York City at the start of the project and produced cover maps for the three neighborhoods," said Dave Nowak, project leader for the Forest Service's Northeast Research Station and lead analyst for the Neighborhood Tree Survey Project. In addition, the Forest Service produced larger-scale land-cover maps of the area from the infrared and color 3-foot-resolution aerial photographs.

Source: OASISNYC.net.

Figure 8.7 A map displayed on OASIS shows the location of sixty street trees surveyed in the East Greenwich Village neighborhood in southern Manhattan.

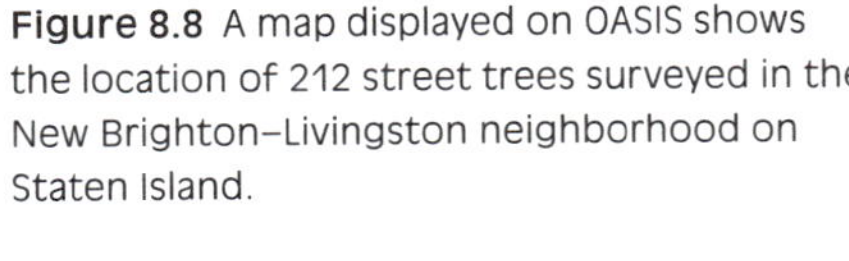

Source: OASISNYC.net.

Figure 8.8 A map displayed on OASIS shows the location of 212 street trees surveyed in the New Brighton–Livingston neighborhood on Staten Island.

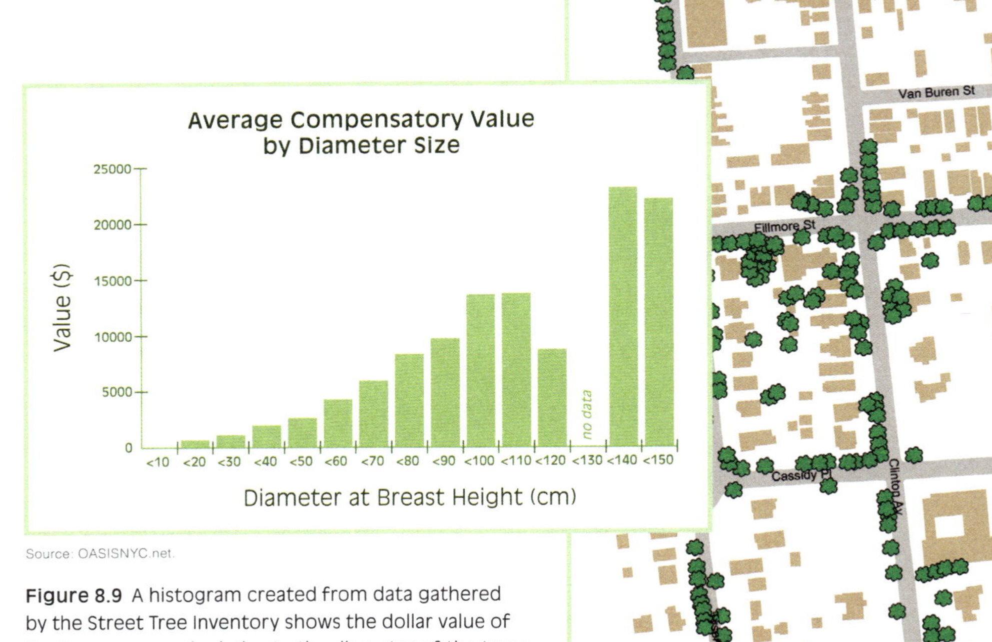

Source: OASISNYC.net.

Figure 8.9 A histogram created from data gathered by the Street Tree Inventory shows the dollar value of the trees surveyed relative to the diameter of the trees.

Source: OASISNYC.net.

Using data collected by the Citizen Pruners, Forest Service researchers estimated the carbon storage values, air pollution filtering, and dollar values of the trees. "Because this was all GIS-based and a tree inventory, we did the analysis tree-by-tree so there would be a link back to the GIS database," Nowak said. Data on population density, tree stocking levels, and tree cover per capita also helped identify areas of the city needing more trees.

"We overlaid population data with tree cover data to create a [tree] planting index," Nowak said. "The purpose of the index was to show areas with a relatively high need for trees. Areas with high population density and low tree cover had high index values indicating where more trees need to be planted." The researchers used ArcInfo software to make maps from the data collected on the three pilot project survey neighborhoods and then conveyed the information to the public through the OASIS Web site using ArcIMS software.

Figure 8.10 A map displayed on OASIS shows the location of fifty street trees surveyed in the Hunts Point neighborhood of the Bronx.

Going public

Next, the Forest Service's analysis was used to create a new map layer for OASIS. The new OASIS interactive map allowed users to click individual trees and find the values for that tree, including species, height, and trunk diameter, as well as carbon storage values and tree replacement cost.

The Neighborhood Tree Survey is about much more than just using GIS and Internet technology to collect, analyze, and display information about trees. Projects like the survey are helping to make connections between communities, private organizations like NYPIRG, and public agencies like the U.S. Forest Service and the City of New York, according to Matt Arnn, a New York City–based landscape architect directing the Forest Service's Metropolitan Initiative.

The Tree Survey, OASIS, and similar projects are helping the Forest Service and other government agencies determine what kinds of green and open space resources exist within cities, place a value on those resources, and get that information to key decision makers in an easily understandable format. Disseminating this information over OASIS to a million-plus viewers per year is what made this project special, Arnn said. "It gives people a new way of thinking about their green infrastructure and their tree resources. The main goal is to get people to care about trees and think of their street trees as more than just street furniture," he said.

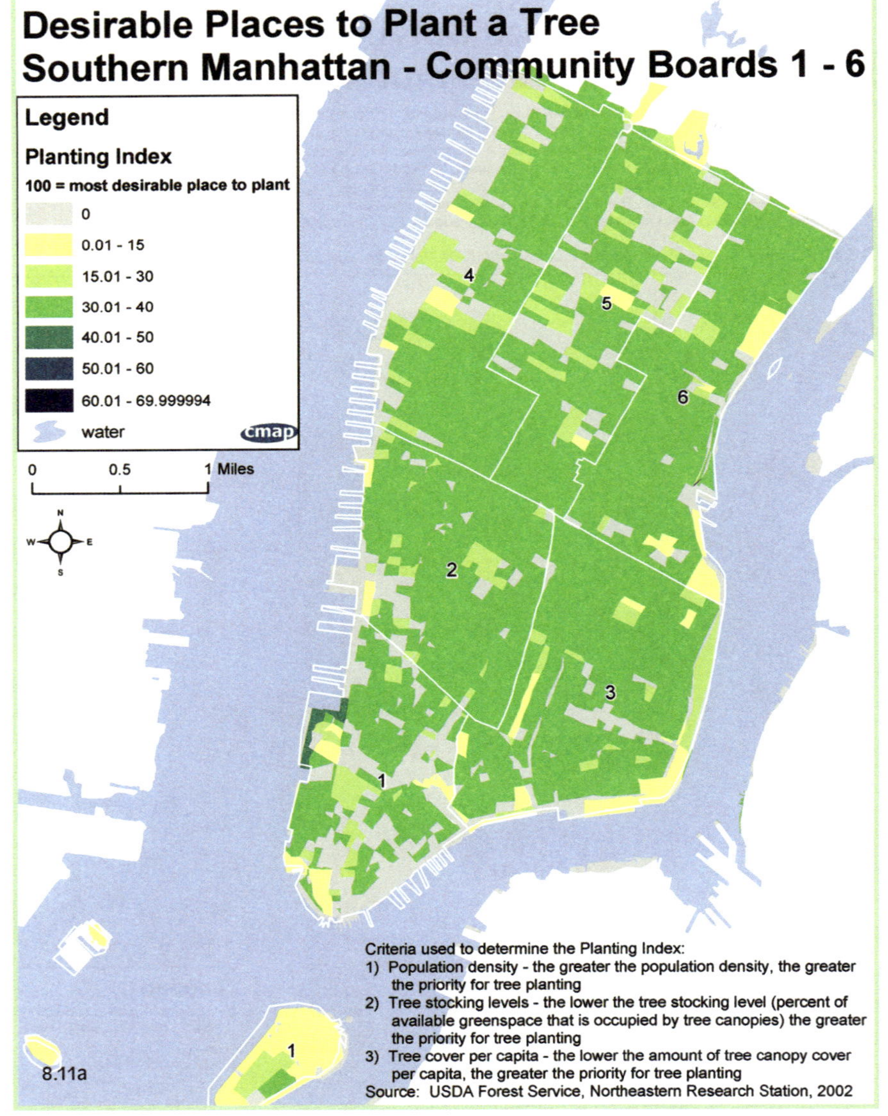

Source: OASISNYC.net.

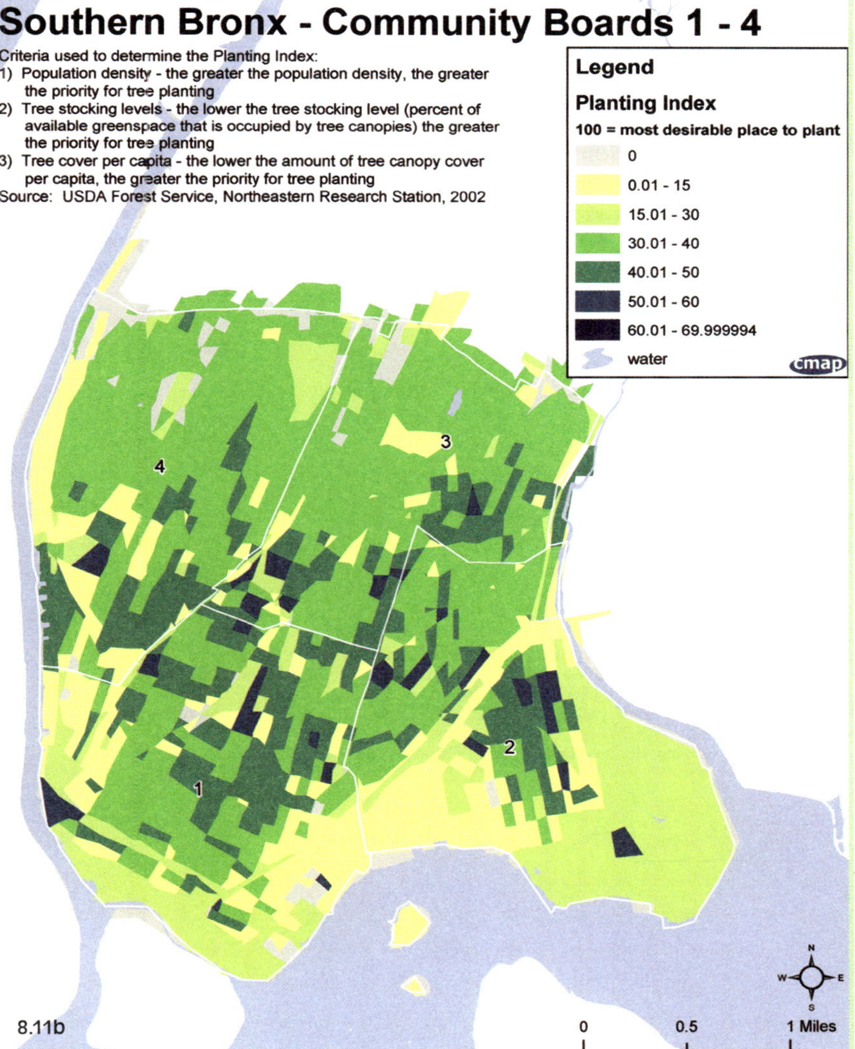

Source: OASISNYC.net.

Another goal was showing that new technologies such as the Internet and GIS enable people to gather information and then present it in useful ways to the public and government. The Neighborhood Tree Survey's pilot program was a success because, with the help of GIS, it helped convince New York City's government officials that collecting information on the city's trees was important, according to Arnn. The survey simultaneously proved the value of GIS-assisted urban tree inventories and developed technologies and techniques that are useful to other cities.

The Neighborhood Tree Survey pilot project opened the door for Trees Count, a new, larger New York City tree survey administered by the city's Department of Parks and Recreation (DPR). Beginning in mid-2005, the DPR conducted a volunteer-driven citywide street tree inventory, but did not collect the same amount of detailed information gathered by the pilot program. Field work for the Trees Count Survey was to conclude in October 2006 with the data used for managing New York's urban forest.

Figure 8.11a, 8.11b, and 8.11c Maps of the most desirable areas to plant new trees in southern Manhattan, the Bronx, and Staten Island. These maps, displayed on OASIS, were created from an analysis of the Neighborhood Tree Survey data conducted by the U.S. Forest Service's Northeastern Research Station.

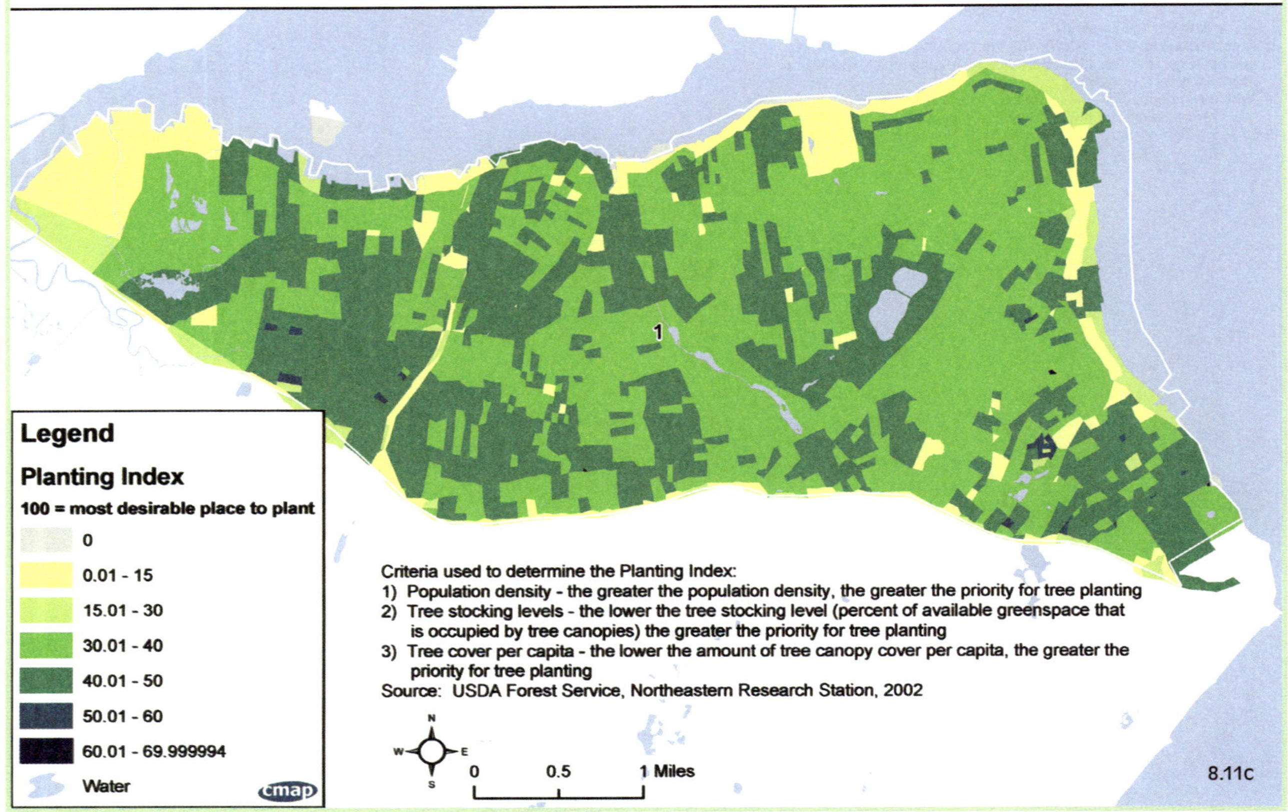

Source: OASISNYC.net.

Tree Inventory 101

After the tree survey pilot program's findings were published in late 2003, Librizzi kept the program alive as an educational opportunity. He worked with high school and elementary teachers and students as well as community groups in Staten Island, Manhattan, and Brooklyn, teaching them how to measure trees and gather data. Librizzi tries to relate it to students' science, math, and geography studies, giving them an introduction to data collection. "Just the act of measuring a tree means that you have to go out and actually put your arms around the tree and touch it, look at it closely, and not just view it from afar," Librizzi said. "That in and of itself has value; it's engaging someone in urban forestry."

Photo by Lenny Librizzi, Council on the Environment of New York City.

Figure 8.12 A student from PS 45 on Staten Island measures the diameter of a street tree. The Neighborhood Tree Survey is an educational program that teaches urban school children about trees while also incorporating other elements of their school curriculum, such as math and science.

References

Librizzi, Lenny. 2004. Neighborhood tree project at PS 45–2004. Professional paper.

New York City Open Accessible Space Information System Cooperative (OASIS), Neighborhood tree survey results. www.oasisnyc.net/OASISTree-default.asp

Romalewski, Steven, Lenny Librizzi, and Mat Cahill, with editing from David Nowak. 2003. Final report to the USDA Forest Service for the urban canopy enhancements through interactive mapping project in New York City.

Trees New York. www.treesny.com

Werner, Jan E. Bisco, Timothy Chandler, Margarte O'Gorman, Jonathan Raser. 2003. Trees mean business: A study of the economic impacts of trees and forests in the commercial districts of New York City and New Jersey. Trees New York and Trees New Jersey, the National Urban and Community Forestry Advisory Council.

Wildlife management 9

Saving birds of prey in northern California

The forests of northern California are home to two endangered raptor species, the northern spotted owl and the northern goshawk.

Data provided by 2004 ESRI Data & Maps.

How do you build scientifically useful databases that can map endangered raptor populations while also producing documentation that will fly in court? GIS is the answer.

Dave LaPlante, under contract with the U.S. Fish and Wildlife Service (FWS) in Northern California, created GIS data management systems for tracking territorial raptors, including the northern goshawk and the northern spotted owl. LaPlante, a GIS analyst and database developer, built the northern Interior Spotted Owl Geodatabase System while working with the environmental consulting firm Resource Management and the nonprofit Northern California Resource Center, based in Siskiyou County.

Accurate data on the spotted owl population is crucial because it affects federal and private resource management decisions that determine when and where to allow timber harvests. Because these issues are so important, the Northern Interior Spotted Owl Geodatabase is designed to produce scientifically sound and legally defensible data that meets the demands of federal regulations. "In order to conserve and recover these species, we need to be able to manage a lot of information in terms of regulatory

Figure 9.1 A sunset lights up Mount Shasta, looking south from the northern end of the Shasta Valley in northern California. The area surrounding Mount Shasta provides habitat for the northern spotted owl and the northern goshawk.

Source: Dave LaPlante, Natural Resources Geospatial.

requirements, forest management projects, and ongoing monitoring," said Brian Woodbridge, supervisory wildlife biologist for the U.S. Fish and Wildlife Service in the northern California city of Yreka, located near the Oregon border.

The Yreka office is responsible for reviewing logging plans of private timberland owners and federal timber sales in inland northern California as they affect endangered species in the region. One of the area's best-known endangered species is *Strix occidentalis caurina,* commonly known as the northern spotted owl. All timber harvest plans in California must be reviewed for their potential impact on the spotted owl using, in part, information from the Biological Information and Observation System (BIOS), the state's official spotted owl database. Gordon Gould of the Wildlife and Habitat Data Analysis Branch of the California Department of Fish and Game developed and manages the database. The BIOS database contains information on spotted owl territory locations and annual survey results, but lacks information to analyze impacts on the bird's habitat caused by vegetation management projects.

Source: Brian Woodbridge.

Figure 9.2 An adult northern spotted owl perched on a California redwood. It is one of the most well-known endangered raptor species.

Figure 9.3 An adult northern goshawk is held for banding.

Source: Brian Woodbridge.

Building a better raptor database

In 2003, LaPlante, the GIS specialist, began work on a northern spotted owl territory-tracking GIS based upon California's BIOS database. The Fish and Wildlife Service wanted to use the state's data in an analytical environment to provide its regulatory biologists with tools to investigate land-use practices and ownership patterns relative to specific territories. The system also needed a flexible format for rapidly updating the GIS database. "Given the nature of the periodic release of the BIOS data, we had to be sure we could automate to the greatest degree possible the reprocessing of those data and the production of the information products that come out," LaPlante said. Records for every year from 1971 to present track the activities of several hundred owl territories.

Until the GIS was created, the U.S. Fish and Wildlife Service didn't have a system for quickly and accurately accessing years of spotted owl monitoring and habitat data.

"In the past we would have to do a lot of scrambling," Woodbridge said. Before GIS, when landowners proposed a timber harvest project, the FWS would consult the BIOS data and past harvest plans, looking for data on territories located within the proposed project's area. The next step was compiling a wide range of records associated with the project site, with sources ranging from computerized data to handwritten field observations. "We would spend a lot of our time just pulling together information, and we weren't always that effective at it," he said. "It's kind of hard to keep that much complex information at your fingertips."

The GIS gathers all of the scattered historical records and references them spatially. Biologists can conduct analyses and track the health of endangered species and their habitats in ways that were previously impossible. "With this database as a tool we can look at all of these things simultaneously in a spatial context," Woodbridge said. "It allows us to generate reports on the status of territories and their habitat through time."

The first step was getting the BIOS information into the system, developing the geoprocessing workflow methodologies and the reporting subsystem. They also created basic location information—township, range, and section and quad maps—refining the structure of the data slightly, according to LaPlante.

Managing the state's original BIOS data takes place within two nonspatial tables, one for observations and the other for bird territories. The BIOS data runs through a preprocessor that redefines certain data fields and creates a couple of initial overlays that capture information such as quadrangle maps associated with the territories. Once the preprocessing is complete, the tabular data is loaded into the geodatabase. The geodatabase then creates spatial features representing the locations of the spotted owl territories.

The next step is to generate three individual buffers at various distances from the center of each owl territory. The buffers help create overlays depicting several aspects of the owl's activities, from the full range of the owl's activity for the largest buffer to the core nesting area for the smallest buffer. After processing the basic features, LaPlante creates a series of relationship classes between activity center features, observation records, and the derived buffers, leveraging the power of relational data management as implemented in the geodatabase.

Northern Spotted Owl Management Geodatabase workflow and process shown in six steps

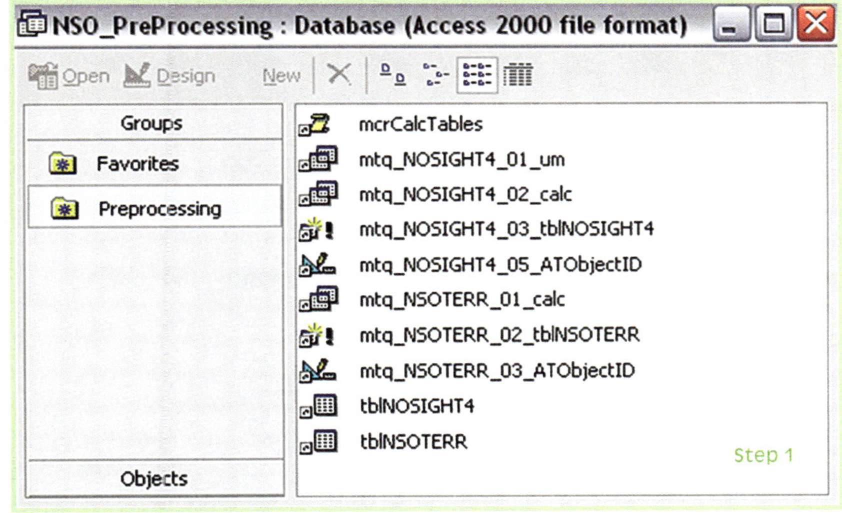

Source: Dave LaPlante, Natural Resources Geospatial.

Figure 9.4a Step 1. Official state spotted owl data provided by the California Department of Fish and Game is initially preprocessed through a Microsoft Access SQL system for validation, transformation, and calculation of required fields.

Figure 9.4b Step 2. An ArcGIS geoprocessing model is used to load the geodatabase with all feature, object, and relationship classes derived from the state data. These include activity center points derived from x- and y-coordinates, management-related buffers at 0.25-mile, 0.7-mile, and 1.3-mile radii used for various overlays, and relationships between historic observation and activity center records.

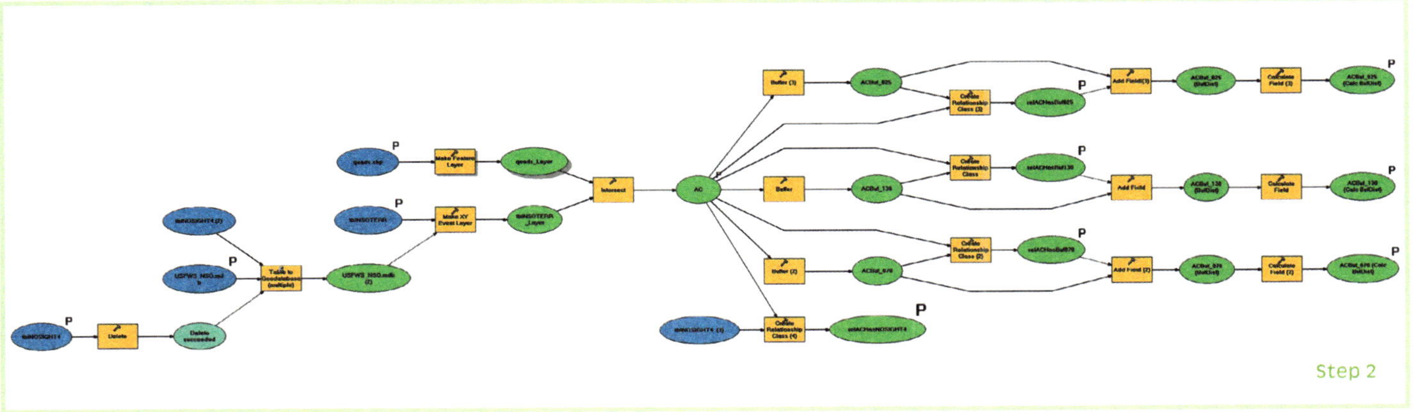

Source: Dave LaPlante, Natural Resources Geospatial.

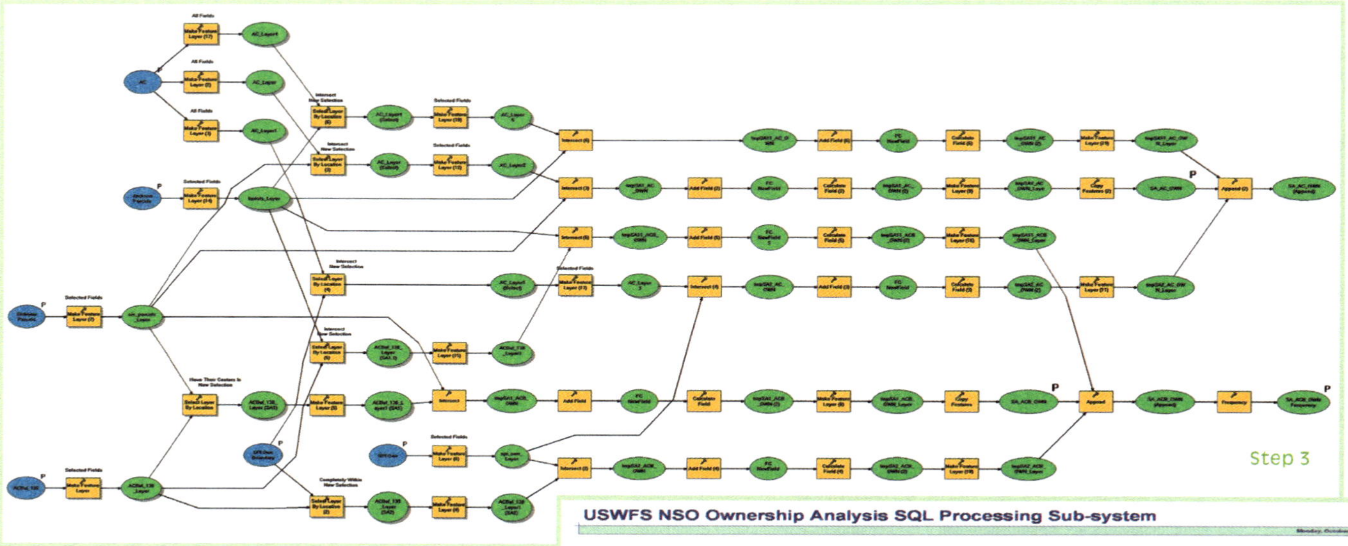

Step 3

Source: Dave LaPlante, Natural Resources Geospatial.

Figure 9.4c Step 3. A second geoprocessing model is used to overlay management buffers with ownership and land-use information to derive the necessary data for analysis.

Figure 9.4d Step 4. A Microsoft Access SQL subsystem postprocesses data developed using the preceding geoprocessing models to calculate results in tables defining (1) ownership and land-use patterns for each activity center, and (2) an annual reproductive status evaluation derived from historical observation or survey records.

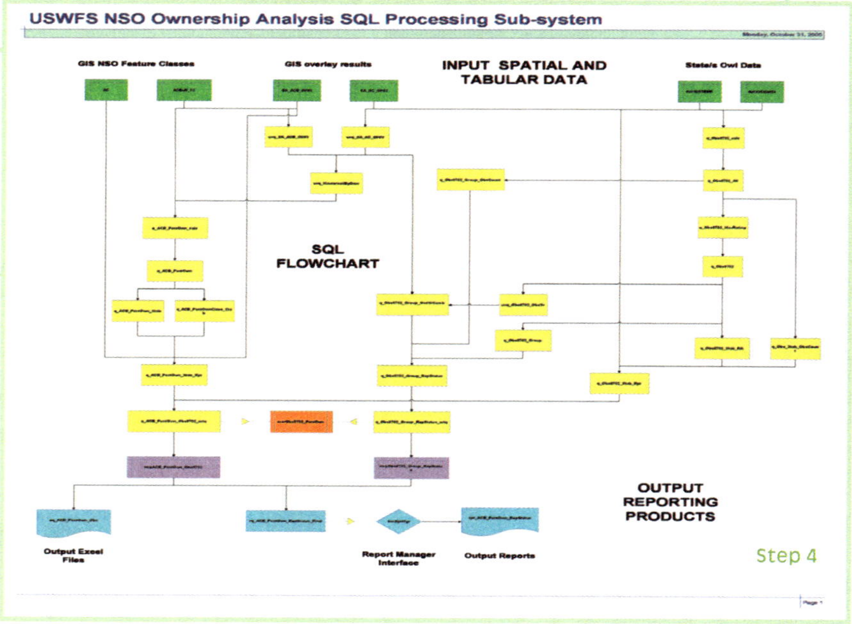

USWFS NSO Ownership Analysis SQL Processing Sub-system

GIS NSO Feature Classes

GIS overlay results

INPUT SPATIAL AND TABULAR DATA

State/s Owl Data

SQL FLOWCHART

OUTPUT REPORTING PRODUCTS

Output Excel Files

Report Manager Interface

Output Reports

Step 4

Source: Dave LaPlante, Natural Resources Geospatial.

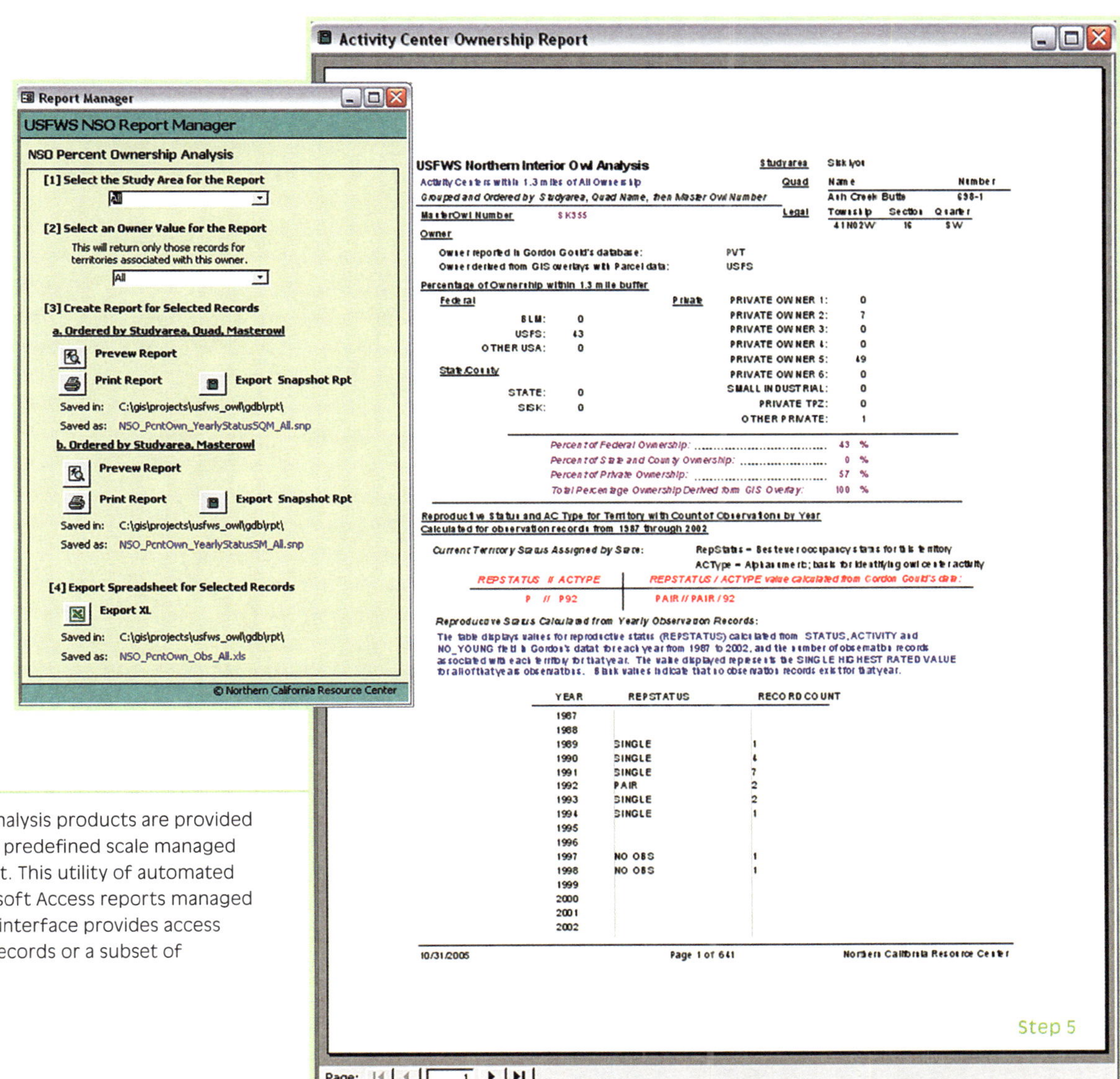

Source: Dave LaPlante, Natural Resources Geospatial.

Figure 9.4e Step 5. Data analysis products are provided through regional maps at a predefined scale managed through a Map Book project. This utility of automated map production and Microsoft Access reports managed through a report manager interface provides access to multiple reports for all records or a subset of processed records.

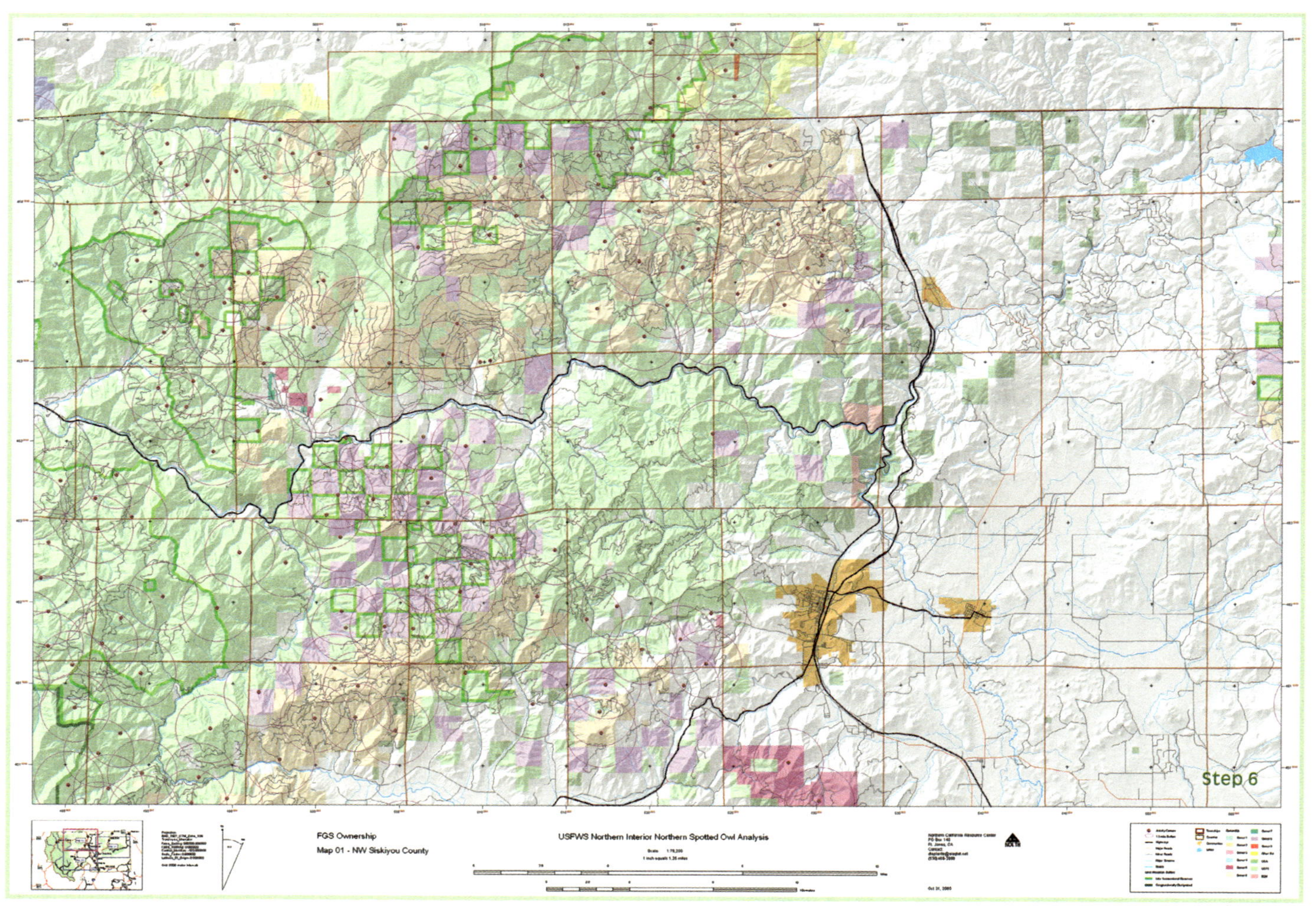

Source: Dave LaPlante, Natural Resources Geospatial.

Figure 9.4f Step 6. Northern spotted owl analysis map produced for the U.S. Fish and Wildlife Service using all aspects of the raptor management database.

Mapping bird territories

The FWS biologists needed maps of the known spotted owl territories within the northern interior of California based on occupancy and reproductive status. The maps are necessary for annual status reports on any given territory. Territories were assigned status classes ranging from individual owls to actively reproducing pairs.

Biologists reviewed the reproductive status of owl territories within the areas proposed for management activities. This was necessary so biologists could identify territories possibly affected by activities such as timber harvests. Evaluating the cumulative effects of past timber harvests to determine the relative impact of a proposed project was also an important function.

GIS allowed for collecting the scattered historical records of territory surveys and management within those territories. It made it possible to digitize and spatially relate the records. Biologists can conduct analysis and track the health of endangered species in ways that they previously could not. "With this database as a tool, we can look at all of these things simultaneously in a spatial way," Woodbridge said. "It allows us to generate reports on the status and habitat condition of owl territories, which dramatically aids us in our work to conserve and recover the species."

Source: Brian Woodbridge.

Figure 9.5 An adult female northern goshawk tends to her nestlings. Determining the location and number of mating pair raptors is one of the most important functions of the raptor management geodatabase.

The reporting function of the GIS, which uses a Microsoft Access report manager interface, is one of the most useful features of the system. Ultimately, this report and its associated map documents are the primary information products of the system. "We needed to distill the results of multiple surveys into annual activity status values that provide the biologist a clear understanding of what has and is occurring at each territory over time, and provide clear visualizations of those results," LaPlante said.

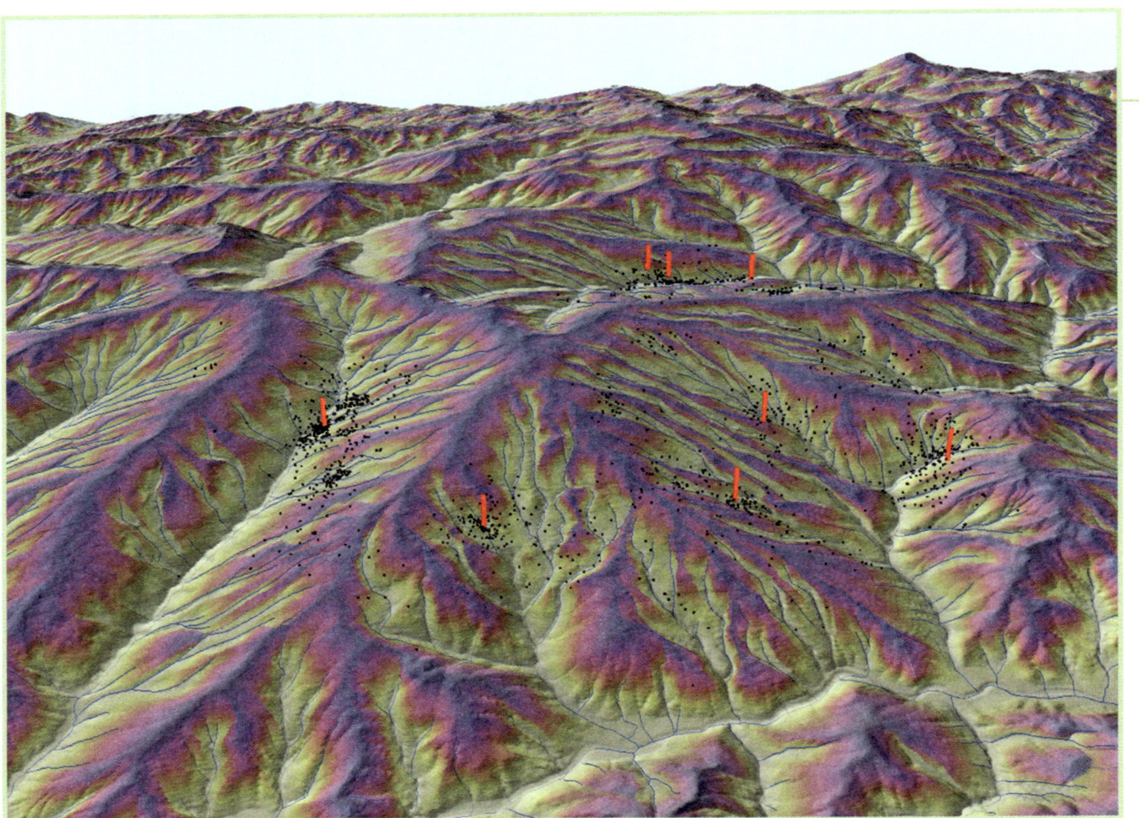

Figure 9.6 Analysis of biotic and abiotic factors has been employed to develop owl management strategies in the region. Of all the abiotic factors assessed for significance, slope position has one of the strongest correlations. The algorithm used defines valley bottoms and ridgelines based on flow accumulation thresholds. The algorithm then calculates a slope position (0–100) for each pixel in between. This landscape is looking north over the Klamath River and displays nest site locations (red bars) and radio telemetry points from a four-year tracking study of local owl populations.

Source: Dave LaPlante, Natural Resources Geospatial.

Rethinking the facts

Evaluating where information didn't exist was a key mission of the database. "Positive data is easy to keep. Negative data, such as surveys that did not detect owls, is seldom effectively maintained and utilized," Woodbridge said. "Now we have a way to manage those data. We can see where surveys have not been conducted and get a much better picture of what's going on in the landscape and where the uncertainty lies."

It was critical to know when and where surveys had taken place without owls being seen. Identifying years when surveys had not been conducted was also very important. Finding gaps in the data is important because it helps the FWS and timber managers tell state regulators what areas lack information. Ownership data built into the system is important because it helps determine who is responsible for managing and surveying territories.

Wildlife biologists are increasingly taking on the task of managing uncertainty. "What we have to be able to do is clearly describe where uncertainty lies and how we made decisions where uncertainty is a factor," Woodbridge explained. "Very often, especially when science gets into the legal realm, there's an expectation of certainty that doesn't exist in our understanding of natural systems."

Using the geodatabase, the FWS biologists can now show areas where surveys have taken place and where they have high certainty of where owls are and are not. GIS also clearly shows the areas not adequately surveyed and where there is less certainty about spotted owl activities. The GIS can also help predict areas where owls are likely to exist. "We can process vegetation data in the GIS and generate reports that show us where we would be more certain that owls might exist based on models of spotted owl habitat selection," Woodbridge said.

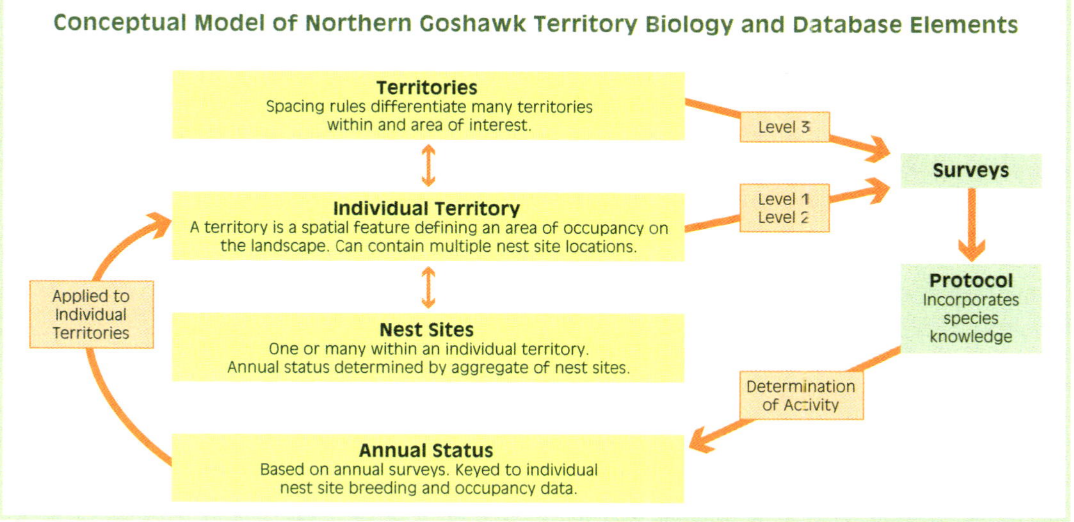

Source: Dave LaPlante, Natural Resources Geospatial.

Figure 9.7 The conceptual model for goshawk biological assessments and framework for database design.

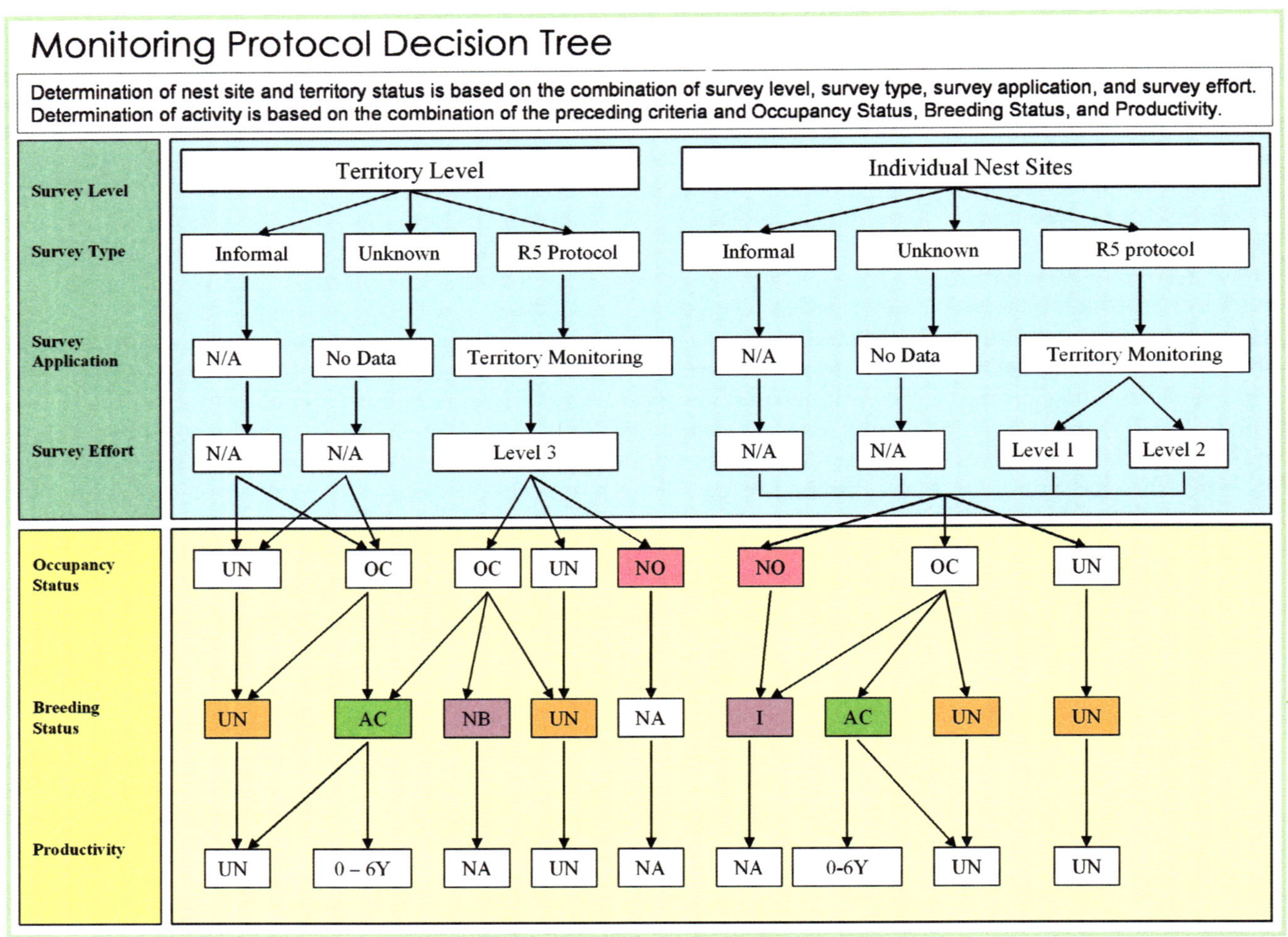

Source: Dave LaPlante, Natural Resources Geospatial

Geoprocessing makes a big difference

New ArcGIS software was released just as LaPlante was beginning to build the spotted owl GIS. The ESRI software allowed LaPlante to fully automate the geoprocessing workflow and data maintenance in the GIS.

Geodatabases are a powerful medium for data management and analysis, and the integration of geoprocessing into ArcGIS made it possible to create a complex automated workflow in the relational database framework. "In my work, 80 percent of what I do is all about databases. Developing in the geodatabase provides the opportunity to integrate GIS with the inherent processing and reporting capabilities of the relational database," LaPlante said.

Before this geoprocessing framework was released, workstations using ARC Macro Language (AML) and coverages performed the automation of geoprocessing. The process required exporting data managed in a geodatabase to coverages, then processing with AMLs, then importing the results back into the geodatabase. "The ArcGIS 9.0 geoprocessing framework allows us to do that directly within the geodatabase, so we can bypass the need to use AML programs and coverages as our automation component for geospatial processing," LaPlante said. "The ability to fully automate geoprocessing within the geodatabase environment is huge—it's the core of our workflow. It's efficient and powerful, and has allowed us to develop complex integrated solutions."

Figure 9.8 The Monitoring Protocol Decision Tree is used to derive a consistent and defensible annual occupancy and reproductive status for surveyed Northern Goshawk territories, which ultimately drive management decisions. The decision tree accounts for the subtleties implicit in survey scale, protocol type and application, level of effort, and observations to derive factors such as annual breeding and occupancy, and status of active, inactive, not breeding, not occupied, or unknown. It also serves as the framework for the CalGos Database evaluation engine.

Automating functions

Working with current data is important to FWS biologists. LaPlante's system consists of multiple independent subsystems that leverage Microsoft Access SQL processing functions with the ArcGIS geoprocessing capabilities to fully automate the flow of information from raw data to reporting and map production. LaPlante used the DSMapBook extension installed as part of the ArcGIS Developer's Kit to create an automated map production element for the system consisting of four regional map documents. This allows individual or batch map document production.

"When updated information is provided by the state, which can happen several times a year, we need to reload the system and reprocess spatial and tabular data efficiently," said LaPlante. "This is how we can provide products to the biologists that are reflective of the current data." One of the major advantages of the geodatabase system is that it can be developed and managed on more powerful systems, like the one LaPlante has in his office, and distributed to more modestly equipped machines.

Goshawk GIS paves the way

The spotted owl management GIS system wasn't LaPlante's first experience developing a data management system for territorial raptors. He developed a northern goshawk system for the U.S. Forest Service, which managed information on about 1,200 goshawk territories in California. The goshawk system was originally designed as a tabular database for supporting a regional conservation assessment of the species. It had no link to spatially derived habitat data. Its key innovation was an evaluation engine derived from an observations-based decision tree.

"Our learning curve was really steep with the goshawk database," Woodbridge said. A very basic database contained a huge amount of goshawk data. Woodbridge and LaPlante worked together to create a logic engine for the goshawk database that essentially questions itself about the quality and classification of data entered. The logic engine evaluates the level of effort expended during multiple surveys. It also controls the conclusions made based on the annual territory status of each bird and the quality or accuracy of the field surveys. A conclusion of "not occupied," for example, could not be drawn from a survey with low effort. The logic engine made entering data that didn't conform to specific quality standards impossible. "It's really important to control the quality of the data when it's being entered, not later during evaluation and reporting," Woodbridge said.

The GIS database systems LaPlante develops differ as much as the animals they study, but all use common techniques and tools. "Every time we look at developing data management systems for a specific species, there are

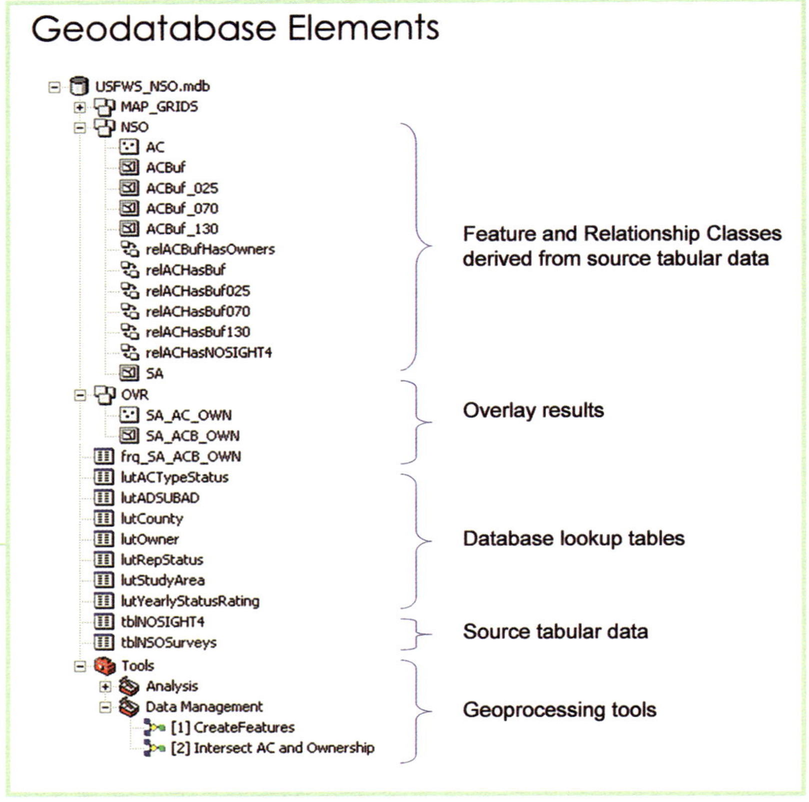

Source: Dave LaPlante, Natural Resources Geospatial.

Figure 9.9 This diagram displays the feature datasets; feature, object, and relationship classes; and geoprocessing tools developed within the Northern Spotted Owl geodatabase. It also demonstrates the functional utility of the ArcGIS 9 framework.

a whole different set of inputs we're going to use and a whole different set of needs and assessments to be derived from those data," LaPlante said. "These wind up being purpose-built systems relative to the needs of the biologists studying each species."

Improving and expanding

The need for high-quality data that meets rigorous scientific standards drove the designs of both systems. "It's always a question of how good the data is you're working with relative to what you derive at the other end," LaPlante said. While each GIS system LaPlante develops is unique, common traits and basic nuts and bolts of the system can have other uses. Techniques developed for automating both spatial and tabular data processing are adaptable for other projects. "I think it's a template for the way that any species-specific information might be used at a management level," LaPlante said.

According to Woodbridge, the current spotted owl database is actually a prototype with a restricted geographic range covering a portion of northern California that represents a few hundred territories. He would like to see the system expanded to cover a wider area with more information about a larger number of owls.

The data management and processing framework developed for the raptor management database may take flight in other applications. "At some point you need a database that has been up and running and used successfully in a number of applications before people are going to see the utility of using and adopting it," Woodbridge said. "That's what's so nice about these geodatabase products. They have pretty broad utility; with just a little tweaking you can go from one species to the next."

Source: Brian Woodbridge.

Figure 9.10 Three-week-old goshawk nestlings. The forest habitat that will help nurture these two downy chicks is rapidly disappearing, a fact that makes the role of GIS in management of the species all the more important.

References

LaPlante, Dave. 2004. GIS management systems for terrestrial raptors. Presented at ESRI User Conference, San Diego, Calif., August 7–13. gis.esri.com/library/userconf/proc04/abstracts/a2034.html

———. 2004. Northern spotted owl management database. Map Gallery poster, ESRI User Conference, San Diego, Calif., August 7–13.

Military use and mitigation 10

Charting unexploded ordnance in West Virginia's wilderness

West Virginia's Dolly Sods region, a beautiful setting that belies the danger lurking below the surface.

Data provided by 2004 ESRI Data & Maps.

The spectacular beauty of the Dolly Sods region in the Monongahela National Forest in West Virginia hides a lurking potential hazard that GIS has helped manage effectively.

Dolly Sods is a rugged 18,000-acre region of uplands and wind-swept plains located on the Allegheny Plateau. The terrain here features flat rocky plains, upland bogs, beaver ponds, and expansive vistas. With elevations of 2,600 feet to more than 4,000 feet, the climate and flora on this high plateau resembles northern Canada. Many plant species in Dolly Sods are typically found much farther north, according to the U.S. Forest Service.

This idyllic setting, however, is the former site of U.S. Army training maneuvers and today, GIS is helping to minimize the risk of unexploded ordnance left behind. Dolly Sods was once covered with a layer of humus up to nine feet thick that formed a magnificent natural carpet under a forest of red spruces and hemlocks. The typical tree was four feet in diameter. Logging, teamed with slash-and-burn agriculture, moved into the area in late 1800s, felling the trees and destroying the fertile humus layer.

During this era, the region acquired its current name. In the early twentieth century, the pioneer Dahle family burned the logged area to create grazing land, or "sods." The family's German surname became the "Dolly" in Dolly Sods. The land was eventually purchased by the Forest Service, and during the 1930s the Civilian Conservation Corps planted conifers and helped build a road, which today provides the main access into the wilderness area.

Photo by David Fattaleh, West Virginia Tourism.

Figure 10.1 The Dolly Sods Wilderness Area, shown here in the fall, has a unique climate and terrain that was ideally suited for training the U.S. Army's XIII Corps for action in the mountainous areas of Europe during World War II.

Dolly Sods gets drafted

At the height of World War II, thousands of soldiers from the U.S. Army occupied Dolly Sods, which was part of the more than two million acres temporarily called the West Virginia Maneuver Area. This friendly invasion was a significant part of the war effort. These troops were training for battle in the mountains of Europe. "The terrain was similar to what they were going to face overseas," said Dr. Erich Guy, a hydrogeologist with the U.S. Army Corps of Engineers (USACE) Huntington District in West Virginia. He manages technical aspects of the Dolly Sods project, which is funded by the Defense Environmental Restoration Program. The unexploded ordnance (UXO) project team includes the USACE, the Forest Service, the West Virginia Department of Environmental Protection, the West Virginia Division of Forestry, and S&C Public Relations.

Known as the Mountain Divisions and famed for their bravery while fighting in the mountains of Italy and Germany, these soldiers used Dolly Sods as a vast classroom for learning how to use artillery, mortars, and heavy machine guns in a mountainous environment. Live fire

Figure 10.2 A. Aubrey Bodine, a photographer for the *Baltimore Sun*, photographed soldiers with the U.S Army's XIII Corps, later known as the Mountain Divisions, learning to mountain climb in preparation for deployment to Europe during World War II.

Photo by A. Aubrey Bodine, © 2005 Jennifer B. Bodine.

and other training exercises took place in Dolly Sods from August 1943 until July 1944.

An estimated 50,000 to 100,000 soldiers streamed through the area during the eleven months of training in the area, according to Robert Whetsell, a historian and filmmaker who researched the U.S. Army's activities in the area for the U.S. Forest Service. Wartime accounts described the nearest town, Elkins, West Virginia, as "a sea of khaki."

Units rotated in, usually for two weeks of training at a time. The Mountain Divisions trained at various sites in this area of eastern West Virginia, but Dolly Sods was the primary artillery range. Munitions fired during the training included 40-mm and 57-mm armor-piercing rounds; 60-mm, 81-mm, 105-mm, and 155-mm high-explosive projectiles; smoke rounds; and 4.2-inch inert (sand-loaded) rounds.

Figure 10.3 Soldiers training in Dolly Sods during World War II fired numerous types of artillery pieces similar to this one. Photojournalist A. Aubrey Bodine captured this image of artillery practice after the war's end.

Beautiful scenery, dangerous duds

The troops fired thousands of rounds in Dolly Sods, but not all of them detonated. In the years following World War II, people continued hunting, fishing, and hiking in the area. In 1975, the 10,215-acre Dolly Sods Wilderness Area was created by an act of Congress to preserve and protect the region and its special opportunities for solitude and recreation, as well as scientific, educational, scenic, and historical values. During 1992 and 1993, the U.S. Forest Service purchased 6,168.5 acres north of the Dolly Sods Wilderness and designated them the Dolly Sods North Area. To the immediate east of Dolly Sods Wilderness and North Areas, 2,268 acres along Forest Road 75 are designated a national scenic area. Although it is remote, Dolly Sods has magnetic beauty that draws 45,000 to 76,000 adventurous hikers, mountain bikers, anglers, hunters, and berry pickers each year.

Figure 10.4 Beautiful and remote, the Dolly Sods Wilderness Area attracts approximately 45,000 to 76,000 visitors annually. Warning signs posted throughout the area alert visitors to the danger of unexploded shells.

Photo by David Fattaleh, West Virginia Tourism. Warning sign image from the U.S. Army Corps of Engineers, Huntington District.

"I think it's one of the nicest places to go camping, hiking, and do outdoor activities in the eastern United States," Guy said. Because of the area's popularity, there have been several efforts to clear the area of unexploded ordnance. In May 1946, a UXO removal team found and destroyed about two hundred rounds of various types, and in 1953 another team found six live rounds.

In 1991, USACE fieldwork estimated the extent of UXO in the Dolly Sods region. The team found thirteen pieces of UXO and nine fragments. Two major UXO sweeps of Dolly Sods conducted in 1997 and 1998 cleared 55 miles of trails to a depth of 1 foot below the surface. These two sweeps also surveyed 178 camping sites and six cabins to a depth 4 feet below the surface. More than 54,000 documented digs were conducted during these two clearances, leading to the recovery of twenty-two live mortar rounds, nineteen inert rounds, and 1,100 pounds of ordnance-related scrap, mostly shrapnel from exploded rounds.

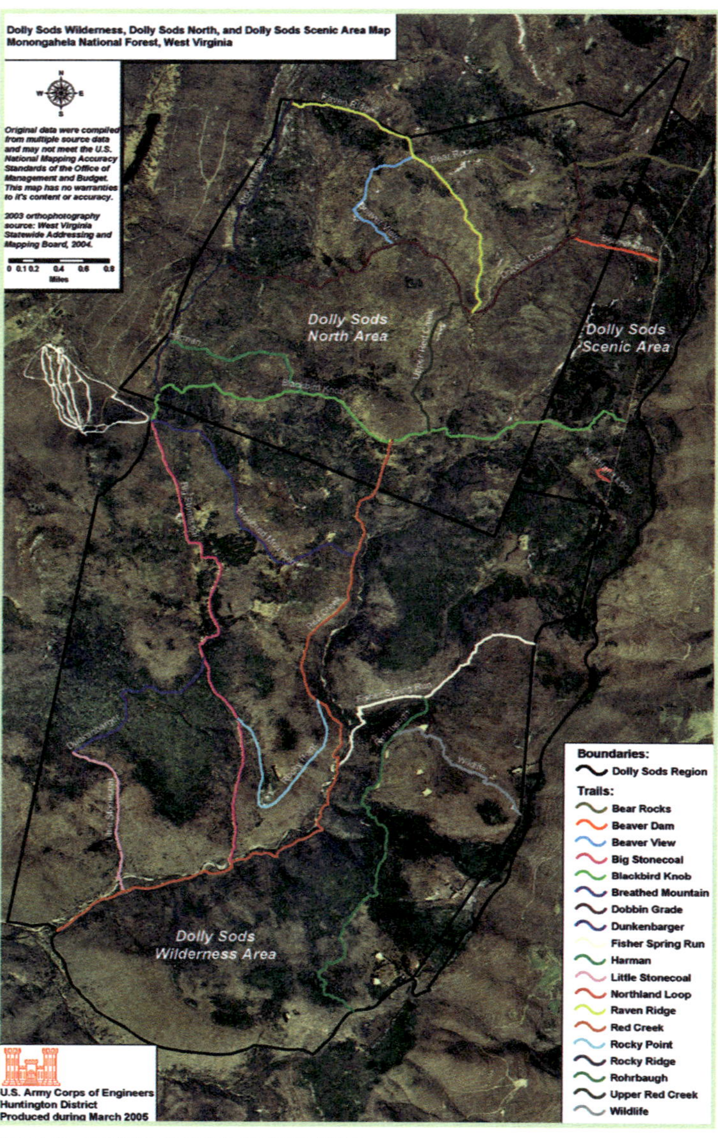

Source: U.S. Army Corps of Engineers, Huntington District.

Figure 10.5 A two-dimensional map with a natural-color, high-resolution aerial photo as a base showing the trails and boundaries of the survey areas used in the U.S. Army Corps of Engineers' unexploded ordnance sweeps. The Dolly Sods region includes the Dolly Sods Wilderness Area, the Dolly Sods North Area, and the Dolly Sods Scenic Area.

Prior removal actions significantly reduced UXO-related risk in the most highly used areas of the Dolly Sods region. The USACE decided GIS was a good solution for managing and determining UXO risks, according to Guy, who manages the project's GIS and mapping aspects.

"My role in the project has been looking at ways to manage risk in a situation where we can't take the risk to zero," Guy said. "The reasons we can't reduce the risk to zero are that the terrain is very rough, and current technology cannot detect all possible ordnance with absolute certainty." Attempting to clear the entire Dolly Sods region for ordnance would be extremely disruptive and damaging to the pristine environment that the U.S. Corps of Engineers and the Forest Service are trying to preserve. "We don't want to deprive the public of the use of these lands, but we do take certain precautions to ensure people are well aware and educated so they remain safe," Guy said.

To manage residual risk without depriving the public from using the lands for their intended purposes, the USACE continues to implement an innovative and extensive public education program. Visitors have sporadically found UXO in the region since the time of training exercises. However, the amount of found UXO has declined dramatically since the removal actions of the USACE. Emblazoned on signs at all trailheads is a contact number for visitors to call in the unlikely event they find UXO. Authorities quickly respond and handle all UXO situations following already established procedures.

Figure 10.6a An unexploded small shell found buried in the ground in Dolly Sods is shown with a pen to give a sense of size.

Source: U.S. Army Corps of Engineers, Huntington District.

Source: U.S. Army Corps of Engineers, Huntington District.

Figure 10.6b Numerous types of unexploded shells have been found in Dolly Sods. This picture, taken from a U.S. Army Corps of Engineers safety video, shows five different types of explosives.

GIS maps unexploded ordnance

Guy began working on the Dolly Sods UXO GIS database in 2003. The U.S. Forest Service, which was preparing a fire suppression plan for the area, requested some maps of UXO finds. "They wanted to have knowledge of possible remaining ordnance distribution in case they have a forest fire, so they could determine where they should be concerned about sending their people in," Guy said. "Obviously fire and unexploded ordnance don't go too well together."

The Forest Service wanted to learn the locations of all finds along the cleared trails and of other areas of UXO finds. Historical research located all of the past finds, ArcView software plotted the finds, and the Forest Service received the resulting maps. The U.S. Corps of Engineers quickly realized the value of further developing and maintaining the GIS for Dolly Sods.

Creating the trail map has been one of the big successes of this project, Guy said. USACE personnel went into the field with GPS units and aerial photography, and with assistance from the Forest Service developed a new trail map for the region. The Dolly Sods trail map used USGS topographic and stream maps as its base. Alternate later versions of the trail maps have employed high-resolution (2-foot-pixel) natural-color aerial photography (provided by the state of West Virginia) as a base. Added to the base were ESRI road coverage, field GPS coordinates acquired by the USACE survey teams, and locations of landmarks and campgrounds. Guy built the GIS using ArcView software running on a Dell Inspiron computer equipped with a 2.2-GHz Pentium 4 processor, 40-GB hard drive, and 512 MB of RAM.

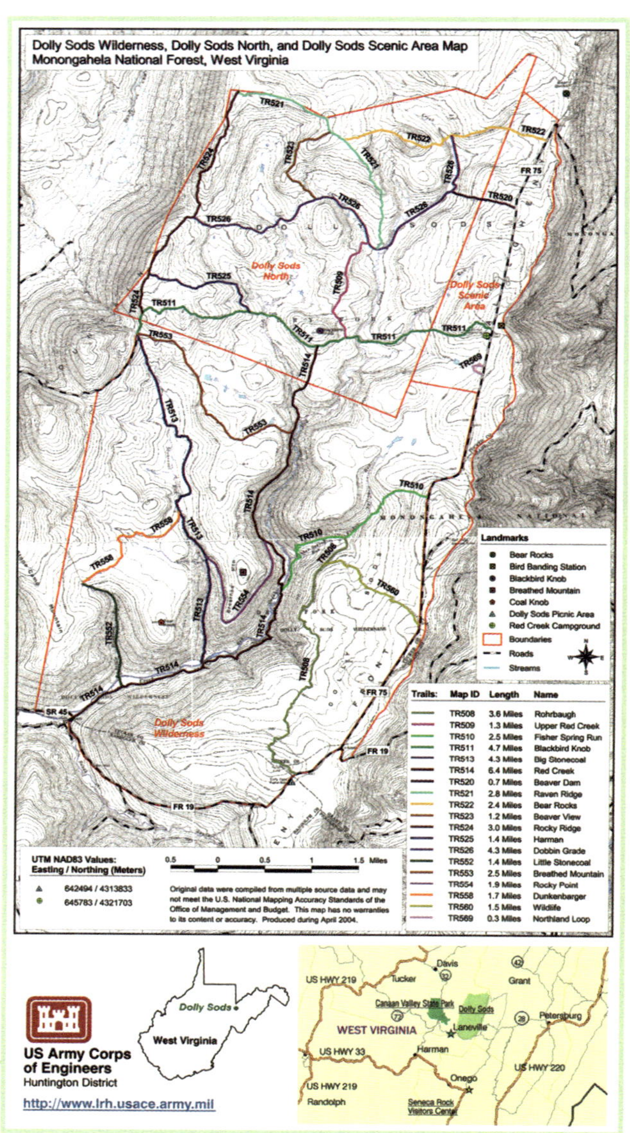

Figure 10.7 This detailed trail map produced by the U.S. Army Corps of Engineers is one of the big successes of the Dolly Sods GIS project.

Source: U.S. Army Corps of Engineers, Huntington District.

New information and continual updates

The Dolly Sods UXO GIS repeatedly proves its utility and is frequently used and regularly updated. "The ability to have information at our fingertips and query it any time we need to is important, and we use it often because a lot of questions come up during our work on the project," Guy said. For example, the Forest Service is currently considering rerouting a trail in the Dolly Sods North Area, and knowing where UXO had been found in this vicinity is important information for safety.

New UXO is found about once or twice per year. Each new find means updating the GIS and creating new maps. Guy also runs other types of queries, such as determining

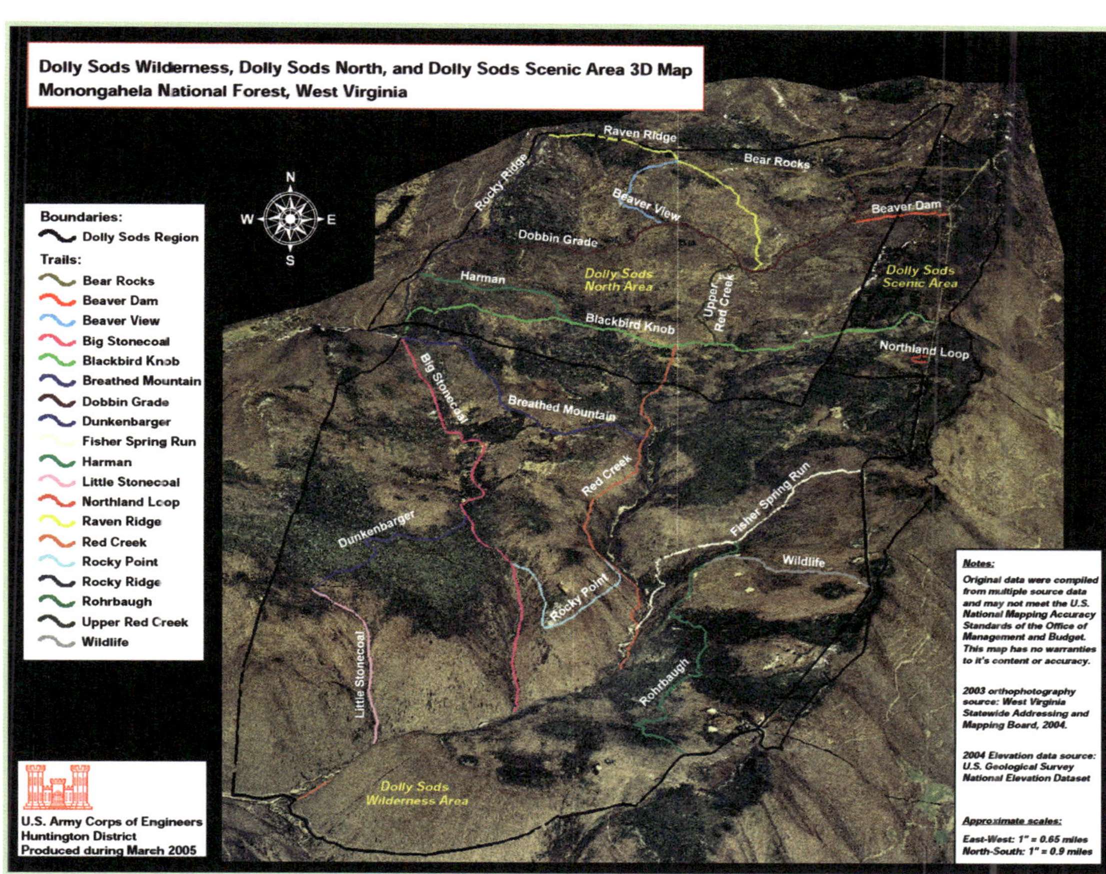

Source: U.S. Army Corps of Engineers, Huntington District.

Figure 10.8 A three-dimensional map of the Dolly Sods survey areas and trails with high-resolution aerial photos as the base was created by the U.S. Army Corps of Engineers for its Dolly Sods UXO GIS project.

the types and locations of UXO finds or the dates of past finds. Georeferenced digital photos of the locations of each UXO find are included in the Dolly Sods UXO GIS database. The UXO sites are photographed every five years and added to the database. The photos create time sequences of site conditions that help document changes in the topography such as any erosion that could potentially expose more UXO.

Guy also continues searching for more historical records of the World War II activity and acquiring more information about past removal efforts. One addition to the GIS was recently discovered documentation concerning the U.S. Army's 1946 and 1953 UXO sweeps. By knowing the ranges of various weapons and ammunition, he determines firing locations, and using the gunners' vantage points he looks for other probable impact zones.

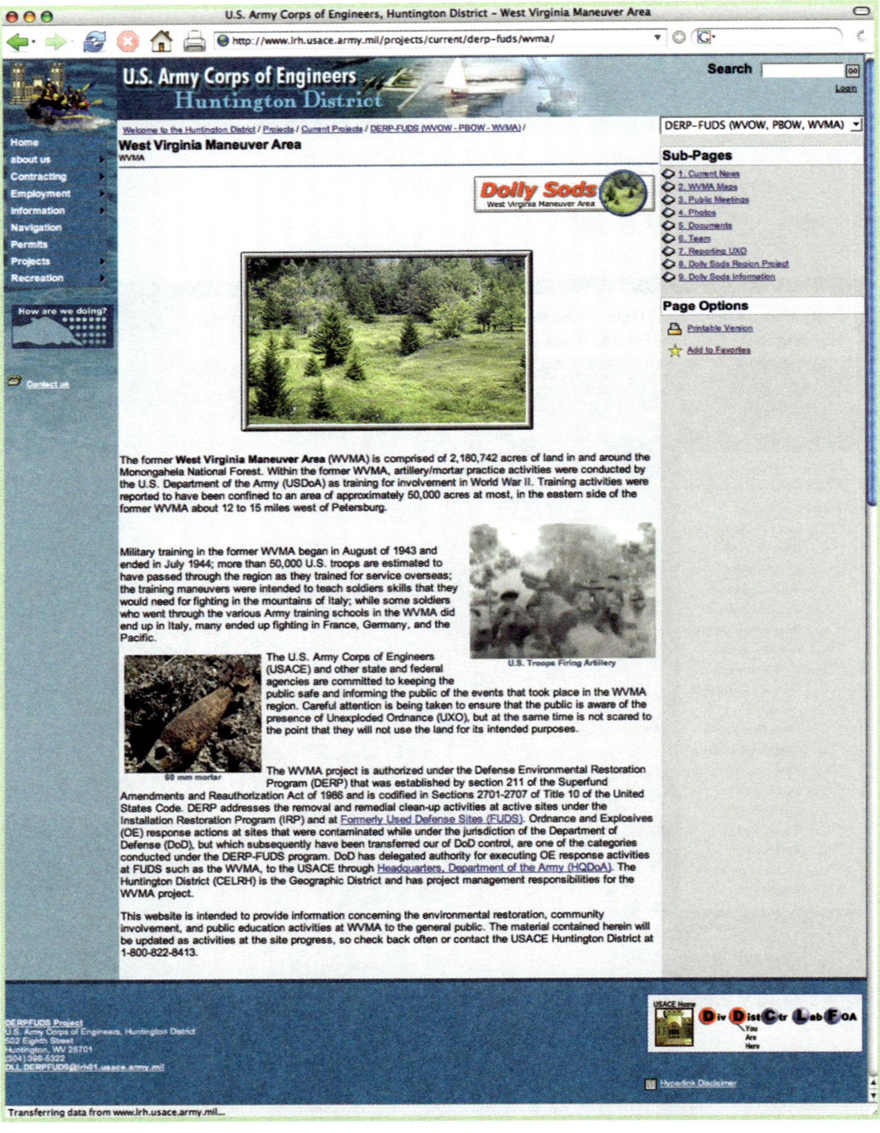

Source: U.S. Army Corps of Engineers, Huntington District.

Figure 10.9 The U.S. Army Corps of Engineers created a Web site to inform the public about the dangers of unexploded ordnance in Dolly Sods. The Web site can be found at www.scpr.com/dollysods/index.htm.

A recently developed Web site informs the public about the U.S. Corps of Engineers' UXO documentation and removal efforts within the former West Virginia Maneuver Area in Dolly Sods. The Web site includes maps, photos, and information about what to do if Dolly Sods visitors encounter UXO.

Amazingly, in the years since Dolly Sods was a firing range, only one injury has occurred from an old shell explosion. In December 1951, while walking in the Dolly Sods with some friends, teenager Wallace Dean was hurt when one of his companions picked up a piece of live UXO and then set it down, according to an informational video produced by the U.S. Corps of Engineers. The device exploded, severely damaging Dean's legs. Fortunately, Dean received prompt medical attention and within a year was able to walk again. Today, Dean is an employee of the USACE in its Huntington, West Virginia, office, and he talks about his experience in the video. Guy said the fact that no one has been injured since that time is a testament to the effectiveness of the UXO removal and public education program.

Rick Meadows, Dolly Sods project manager with the U.S. Army Corps of Engineers Huntington District, also contributed his expertise in the preparation of this chapter.

Source: U.S. Army Corps of Engineers, Huntington District.

Figure 10.10 The U.S. Army Corps of Engineers produced a ten-minute public education video, *Enjoying Dolly Sods Safely,* about the unexploded ordnance.

Photo by Steve Shaluta, West Virginia Tourism.

Figure 10.11 Sunset in Dolly Sods. Although hundreds of exploded shells have been found in the Dolly Sods region since the end of World War II, only one person has ever been injured. GIS has helped the U.S. Army Corps of Engineers greatly reduce the risk posed by unexploded shells in area.

DUSTRAN predicts dust dispersion from military training exercises

During military training exercises, vehicles kick up large dust clouds that can affect air quality in nearby civilian communities. In 2001, the U.S. Department of Energy's Pacific Northwest Laboratory in Richland, Washington, began a four-year project to create an atmospheric dispersion modeling system known as DUSTRAN (an acronym for dust transport) for the U.S. Department of Defense. The purpose of DUSTRAN is to assess the impact of military training and testing on local and regional air quality, manage dust-generating activities, and develop dust-mitigation strategies. The primary

Figure 10.13 ArcMap main window showing the DUSTRAN main control window with example line sources highlighted in the map display.

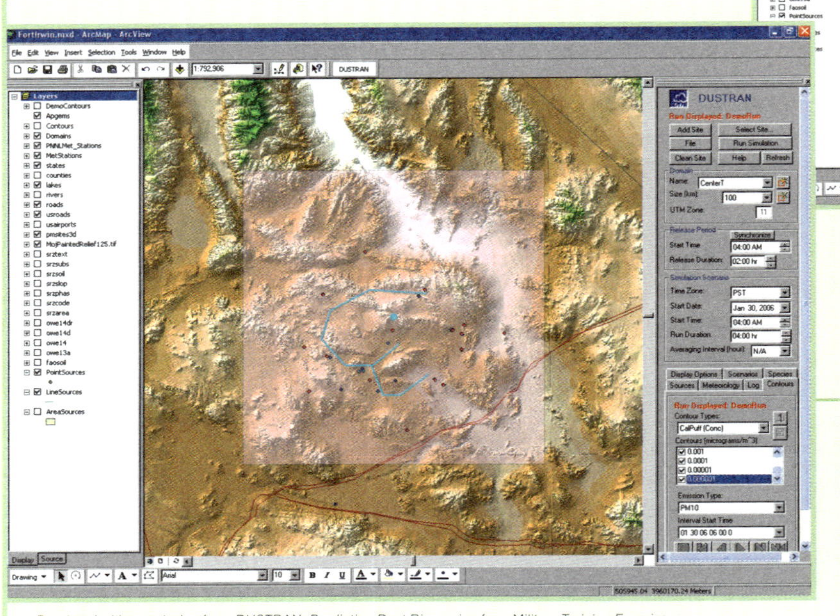

Figure 10.12 The ArcMap main window showing hour-averaged PM_{10} concentration contours one hour after the start of a simulation with vehicle activity on three roads. PM_{10} is a measure of particulate matter—dust—which is smaller than 10 micrometers. This type of dust can settle into the lungs and bronchial tubes and may cause health problems.

users of the DUSTRAN system are officials of military bases where windblown dust is a concern.

The DUSTRAN graphical interface allows its users to identify the location of training activities on a map and to specify the types, speeds, and number of vehicles involved as well as weather conditions such as winds and humidity. DUSTRAN simulates the dust generated by wheeled vehicles. Future versions will include simulations of the dust raised by tanks and helicopters. Users can also specify the time and duration of the training exercise, size of the modeling domain, and duration of the simulation. DUSTRAN can take information directly from surface weather stations or use historical meteorological data.

DUSTRAN users can view the impacts of training exercises from small- or large-scale perspectives. The DUSTRAN dispersion model calculates dust concentrations in the air and deposition fields on the ground and graphically displays within the GIS animations to show a dust plume's movement. Completed in 2005, DUSTRAN runs on ArcMap software. It uses the EPA-approved CALPUFF and the widely used CALGRID dispersion models as well as the CALMet meteorological model. DUSTRAN runs on PCs using Microsoft Windows NT, 2000, or XP operating systems.

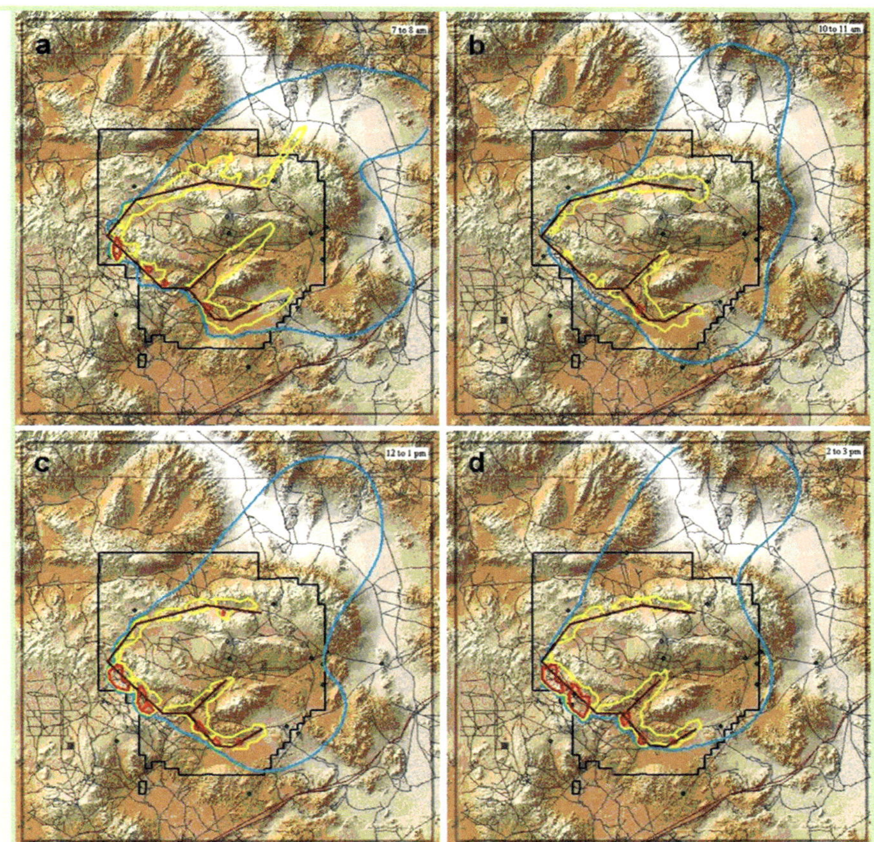

Figure 10.14 Hour-averaged modeled PM_{10} concentrations for training move-out operations. The concentration isopleths are 1 (blue), 10 (yellow), and 100 (red) $\mu g/m^3$. The times shown are ending hour 0800 (a), 1100 (b), 1300 (c), and 1500 PST (d). An isopleth is a line or curve of equal values. Lines on weather maps representing temperature or barometric pressures are examples of isopleths.

One of the biggest challenges in developing DUSTRAN was getting legacy programs to work with the GIS, according to Fred Rutz, a risk and decision sciences engineer at the Pacific Northwest National Laboratory, who helped build DUSTRAN. Rutz and his colleagues developed a dynamic link library that integrated all programs by using ArcMap software, Visual Basic for applications, and ArcObjects software. Tests conducted while DUSTRAN was under development proved very successful, according to a 2004 report by the Pacific Northwest National Laboratory.

DUSTRAN has also attracted the interest of other U.S. government agencies such as the Forest Service and the Environmental Protection Agency. With some modifications, DUSTRAN can also help determine the drift of pesticides sprayed by crop-dusting aircraft. The system could also have other nonmilitary applications. The area around Pendleton, Oregon, has a serious windblown dust problem during the spring months. DUSTRAN may help predict dust storms and calculate dust concentrations in the air around Pendleton.

Figure 10.15 This DUSTRAN map shows hour-averaged modeled PM_{10} concentrations one hour after beginning of training move-out operations with the simulated wind vector field displayed.

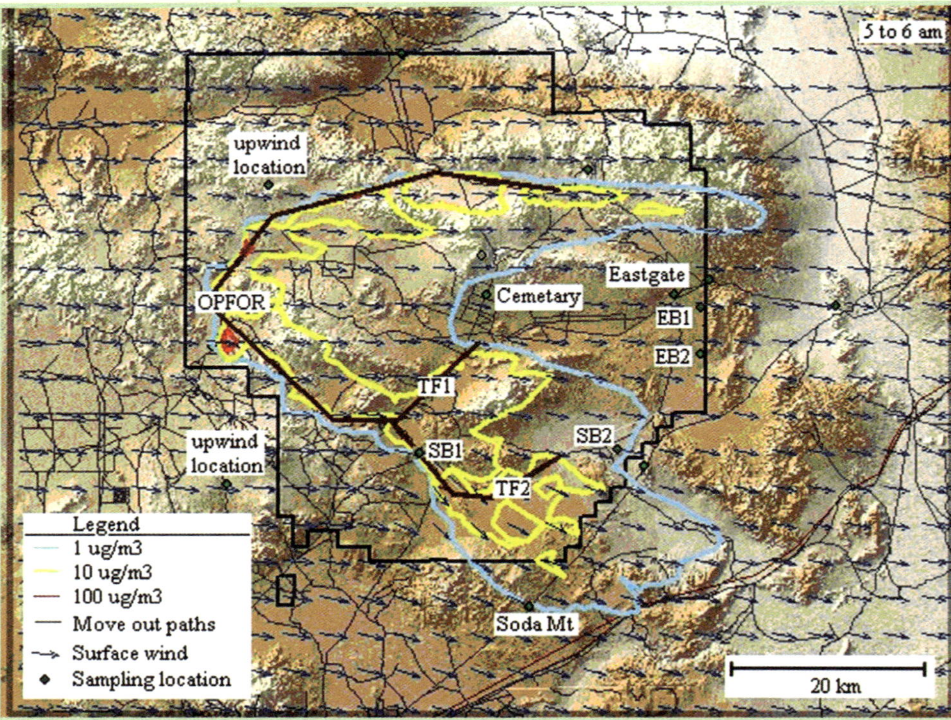

Source: Reprinted with permission from *DUSTRAN: Predicting Dust Dispersion from Military Training Exercises.* Frederick C. Ruts, K. Jerry Allwine. © Battelle Memorial Institute.

References

Allwine, K. J., F. C Rutz, W. J. Shaw, B. G. Fritz, B. L. Hoopes. 2004. Third annual progress report: Development of a GIS-based complex terrain model for atmospheric dust. Prepared for the U.S. Department of Defense Strategic Environmental Research and Development Program under a related services agreement with the U.S. Department of Energy.

Hiltz, Andy. Dolly Sods Wilderness, Monongahela National Forest, West Virginia. The Potomac Appalachian Trail Club. www.patc.net/hiking/destinations/dolysods.html

Ritz, Fredrick, Bonnie Hoopes, Duard Crandall, Kenneth Allwine. 2004. Development of a GIS-based dust dispersion modeling system. ESRI User Conference, San Diego, Calif. gis.esri.com/library/userconf/proc04/docs/pap2054.pdf

U.S. Army Corps of Engineers. 2004. Dolly Sods project, former West Virginia maneuver area, Monongahela National Forest, W.V., ordnance and explosives recurring review, final report.

———. 2004. *Enjoying Dolly Sods safely.* Video produced by S&C Public Relations & Advertising for the U.S. Army Corps of Engineers, Huntington District.

U.S. Forest Service. Welcome to the Monongahela National Forest. www.fs.fed.us/r9/mnf/sp/dolly_sods_wilderness.htm

Whetsell, Robert. 2002. Society of the Military Horse, Web posting. Posting to Web site forum Oct. 10, 2002. www.militaryhorse.org/forum/topic.asp?ARCHIVE=true&TOPIC_ID=1478

Petroleum 11

Cooperating in the Gulf Coast to plan for oil spills

Offshore oil fields along the Gulf of Mexico are a vital energy resource.

Data provided by 2004 ESRI Data & Maps.

The oil tanker *Exxon Valdez* ran aground on a reef in March 1989 in the environmentally pristine Prince William Sound off Alaska's coast, triggering an environmental disaster that drew international attention. With a gash ripped in its hull, the *Valdez* spilled eleven million gallons of crude oil into the sound. Reacting to the Valdez incident, the U.S. Congress passed the Oil Pollution Act of 1990. The legislation gave the Minerals Management Service (MMS), an agency of the U.S. Department of the Interior, increased responsibilities for oil spill contingency planning in marine and coastal environments. The MMS manages U.S. offshore oil and gas resources located on the Outer Continental Shelf (OCS).

Source: *Exxon Valdez* Oil Spill Trustee Council.

Figure 11.1 The oil tanker *Exxon Valdez* ran aground March 24, 1989, in Alaska's environmentally pristine Prince William Sound, spilling nearly eleven million gallons of crude oil. In response, the U.S. Congress enacted a law that gave the Minerals Management Service increased responsibilities for oil spill contingency planning in marine and coastal environments.

No coastal area in the continental United States is more vulnerable to oil spills than the Gulf of Mexico, where a significant portion of the nation's crude oil and natural gas is extracted by offshore platforms or other facilities. The OCS area of the gulf alone accounted for 28.5 percent of all oil production and 19.2 percent of natural gas production in the United States as of June 2005, according the Energy Information Administration. The gulf provides habitats for large numbers of fish, birds, and land animals that live along the coasts and in estuaries. The gulf's fish and animal habitats also attract humans. In Louisiana alone, activities such as bird watching, wildlife photography, fishing, hunting, boating, and the harvesting of alligators and fur-bearing animals accounted for $5.1 billion in sales and supported 77,688 jobs as of 2003, according to a report prepared for the Louisiana Department of Wildlife and Fisheries.

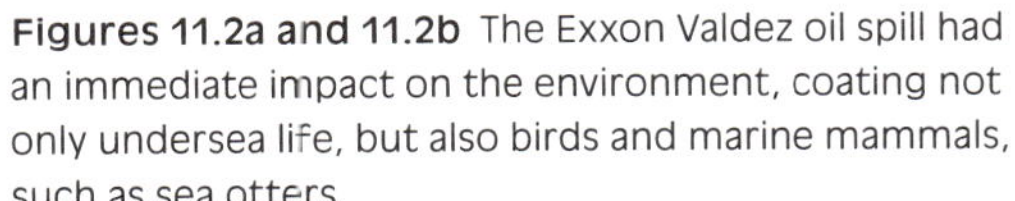

Figures 11.2a and 11.2b The Exxon Valdez oil spill had an immediate impact on the environment, coating not only undersea life, but also birds and marine mammals, such as sea otters.

Overcoming data frustration

Although the Oil Pollution Act required oil companies, along with state and federal agencies, to come up with oil spill response plans, getting everyone to work together proved difficult. In 1997, the MMS began work on the Gulf-Wide Information System (G-WIS, pronounced gee-whiz), a GIS oil spill response planning database covering the entire U.S. portion of the Gulf of Mexico. The MMS, through its Environmental Studies Program, provided more than $2 million to develop G-WIS. Louisiana State University (LSU) served as the coordinator between the various federal agencies in the Gulf Coast states and the oil companies.

G-WIS was born from the frustration of the states, oil companies, and federal government as they tried to develop oil spill response planning tools, according to Dr. Shea Penland, a professor of petroleum geology at New Orleans University. He was at LSU when the G-WIS project began.

While the focus of G-WIS was on an oil spill planning database, it was clear to Norman Froomer, a geographer for the MMS who directed the project, that the data necessary for oil spill planning was essentially the same data MMS required for other kinds of environmental analyses. MMS had been trying to develop an agency-wide environmental database for its corporate database for more than five years without much success when G-WIS came along. G-WIS was an opportunity for MMS to meet Oil Pollution Act needs while advancing its plans for a corporate environmental database.

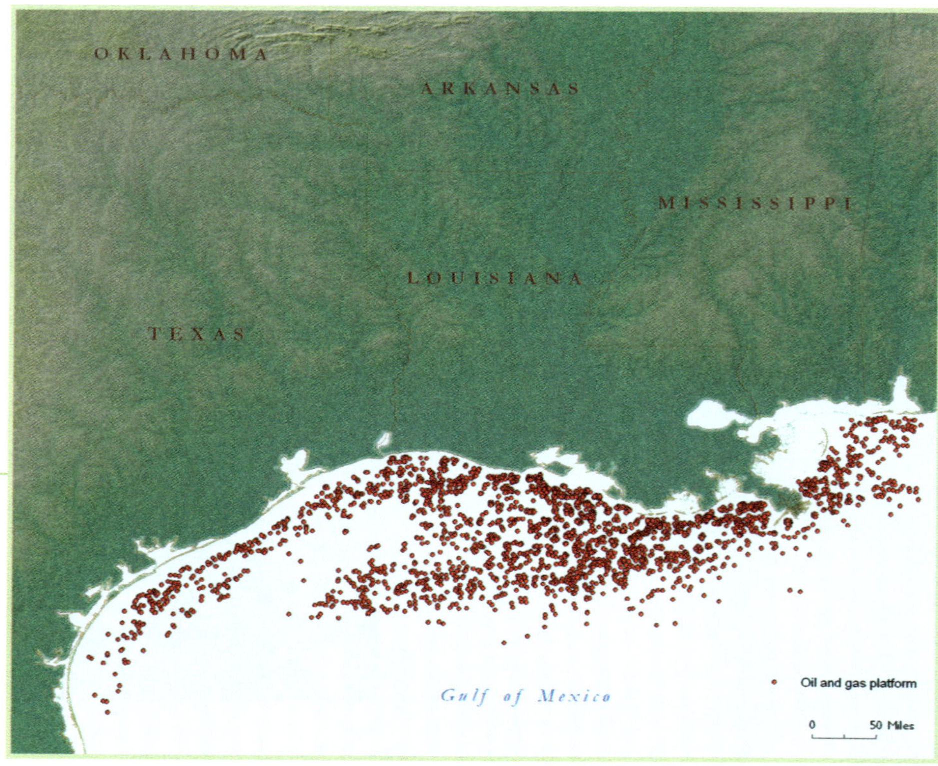

Source: Louisiana Gulf-Wide Information System, Environmental Sensitivity Index and LOSCO Environmental Baseline Inventory Dataset "Oil and Gas Platform Structures in the Gulf of Mexico from MMS source data. Geographic NAD83. LOSCO (1999) [poggeog 3dxmms]."

Figure 11.3 Thousands of offshore oil and gas platforms in the Gulf of Mexico are spread along the coasts of Alabama, Louisiana, Mississippi, and Texas. The large number of platforms and the proximity to beaches and wetlands along the Gulf Coast states illustrates the need for oil spill contingency planning. This map was created from data produced by the Gulf-Wide Information System project.

"It seemed foolish to spend a lot of money and effort on such a complex project that involved so many different people and organizations and wind up in the end with a database only for oil spill planning," Froomer said. "I didn't want to go through the same difficult and lengthy process again when we needed an air quality or some other environmental database."

Under the provisions of the Oil Pollution Act, oil companies were required to submit oil spill response plans to the MMS. The companies would hire consultants to prepare the plans, but the MMS or other agencies involved would often reject them because of deficient data. "I was listening to Clean Gulf Associates and other oil spill cooperatives saying to MMS 'Every time we submit a plan you spit it back to us saying the bugs and the bunnies are wrong,'" Penland said. "They were saying 'Give us the right bugs and bunnies; you tell us where they are and where they nest and that'll be a big deal.' If the regulators had the data, but aren't sharing with the people who are supposed to be using the data, then there was a big disconnect and we had to figure that out." Froomer agreed.

"The oil industry wanted a database that they could use," he said. "They didn't want a database that was OK with the MMS, but the states of Florida or Louisiana didn't like it. They wanted all the regulators—state and federal—to agree to one set of information and that's what we developed."

When Hurricane Katrina devastated a huge swath of the northern Gulf of Mexico in September 2005, the energy industry also suffered. In the hurricane's wake, nearly three-fifths of the region's oil production and one-third of natural gas production shut down because of storm damage. At least eight million gallons of oil was lost in ten large- and medium-sized spills and 134 smaller incidents, according to the U.S. Coast Guard. Numerous federal and state agencies responded to the spills, armed in part with GIS data developed by the G-WIS project.

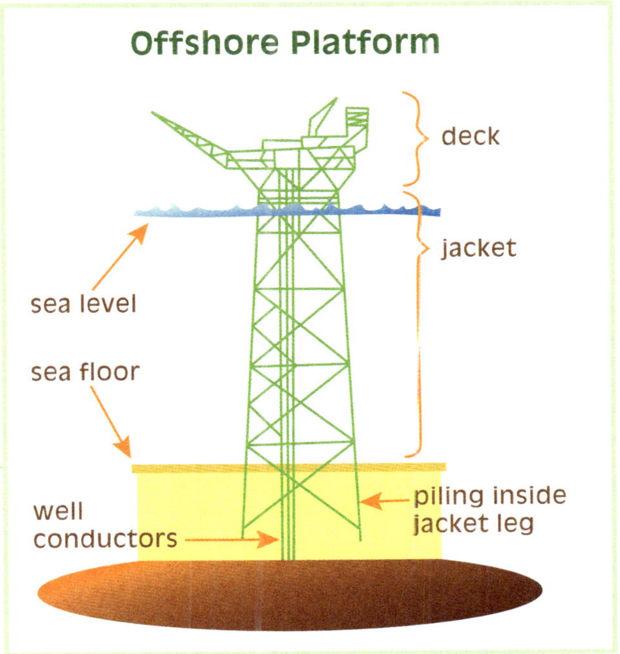

Adapted from Louisiana Gulf-Wide Information System, Environmental Sensitivity Index.

Figure 11.4 Offshore platforms are often massive structures that extend hundreds of feet to the sea floor with drilling and pumping structures that extract oil or natural gas from below the sea bed. Approximately a third of the crude oil and about a fifth of the natural gas produced in the United States comes from the Gulf of Mexico, according to the Energy Information Administration.

Agreeing on data standards

The oil companies and the federal and state agencies quickly began to cooperate with both MMS and industry and chipped in money to fund the program. "It was like a new day," Penland recalled. The oil companies and the Gulf Coast states were able to get the MMS to agree to a one-year moratorium on oil spill response plans. All of the parties involved, except the Texas General Land Office, which did its work in-house, agreed that Research Planning Inc. (RPI) would assemble data from all of the sources and integrate it into a single GIS database. RPI, of Columbia, South Carolina, had been the main contractor for producing the National Oceanic and Atmospheric Administration's (NOAA) Environmental Sensitivity Index (ESI) data and map atlases for more than twenty years.

The Gulf Coast states and the oil industry were familiar with the ESI products and many had worked with RPI in the past. The ESI data structure turned out to be very powerful for coastal and marine environmental analyses, according to Froomer. The G-WIS database used ESI as a foundation. Modifications to ESI accommodated regional data from multiple sources and supported more generic environmental analysis. But agreeing on data standards and building an authoritative GIS was just the beginning of what became a Herculean effort to create the gulfwide database. The technological challenges, while significant, were nothing compared with getting the people involved to work together to decide on data standards.

G-WIS project partners included

- Minerals Management Service
- Louisiana State University
- Louisiana Department of Wildlife and Fisheries
- Louisiana Oil Spill Coordinator's Office
- Texas General Land Office
- Mississippi Department of Environmental Quality
- Alabama Geological Survey
- Florida Department of Environmental Protection
- NOAA Office of Strategic Environmental Assessment
- NOAA Hazmat
- U.S. Geological Survey, Biological Resources Division
- Clean Gulf—an industry oil spill response cooperative

Exxon, Texaco, and other oil companies contributed resources.

Source: Minerals Management Service.

Some of the early meetings for the G-WIS project were difficult, recalled Jim Hanifen, director of the coastal ecology section of the Louisiana Department of Wildlife and Fisheries. Hanifen, who was manager of the Coastal Ecology Program during the time G-WIS was under development, said they found success by going back to basics. "One day we had forty people sitting around a table arguing. We kept throwing out the term 'map' and it occurred to me that we didn't have a common definition," Hanifen said. "I suggested that maybe we needed to work on definitions first. We did, and as soon as we came up with some common definitions, it suddenly all started coming together."

Source: National Oceanic and Atmospheric Administration.

Figure 11.5 Three of the thousands of offshore platforms located in the Gulf of Mexico.

Building a big GIS with a small budget

From the beginning, the G-WIS project faced significant challenges, including the fact that just $2 million had been allocated to create a database covering the entire Gulf of Mexico from Brownsville, Texas, to Tampa, Florida. In addition, agencies from the five Gulf Coast states, four federal agencies, and several major oil companies had to work together while also trying to integrate numerous data sources and formats.

Photo by Dr. Terry McTigue, National Oceanic and Atmospheric Administration, National Ocean Service, Office of Response and Restoration.

Figure 11.6 Boomed marshes containing an oil spill in Dixon Bay, Louisiana, on January 28, 1995.

"For an area of this size, it wasn't feasible to develop new data with this amount of money," Froomer said. A group effort was necessary to create G-WIS. Oil spills don't recognize political boundaries and often span more than one state's territory. The database had to provide consistent information across all data sources and geographic areas, while it also had to speak the same language from a technical standpoint.

"Data was available from states and other federal agencies, but it was all in different formats," Froomer said.

To make G-WIS work, the data from all of the different sources had to fit together into one consistent database, not only with the same projection and file structure, but also in the meaning of the information and how the data was modeled. The federal agencies and the oil and gas industry came to the project with regional and national perspectives because their jurisdictions and operations typically span more than one state. The states, however, had more focused interests. While Florida was somewhat interested in data from adjacent Alabama, data from Louisiana or Texas on the other side of the gulf was another matter. Since the states were providing much of the data that made up G-WIS, it was important for them to support the regional approach, even though they would not benefit directly.

The G-WIS solution was to provide resources for the states to update and format their existing data for internal use but in a consistent format that the federal agencies and the oil industry could use. Froomer found this approach to be a win-win for everyone. Louisiana, for example, was able to convert huge volumes of data from a variety of electronic and paper formats into a consistent database for the state's use. At the same time, MMS and the oil industry could use the Louisiana data for oil spill analyses.

Integrating the numerous datasets began with creating a data dictionary to describe the structure of the data within the GIS, said Lynda Wayne, who was a GIS specialist at Louisiana State University during the development of G-WIS. Building G-WIS required bringing together the oil industry, which had the data on the locations of pipelines, drilling platforms, and other infrastructure, with the natural resources experts, who knew where the fish, birds, and other animals lived. When the project began, Wayne recalled, the G-WIS development team tried to start with a template for all the data with everything done the same way. Mapping the shoreline in Mississippi meant that researchers with GPS backpacks walked the high-tide line. In Louisiana the shoreline is extremely complex with some areas requiring a technical specification to discriminate land from water.

Wayne, now a principal of GeoMaxim, a consulting firm in Asheville, North Carolina, helped develop the data dictionary and the metadata for the project. She also came up with the G-WIS acronym. "The original name was something like 'Oil Spill Contingency Planning Geographic Information System and Database,' and I knew I couldn't go eighteen months to two years living with this long, complicated name," Wayne said. "I started playing with the project parameters and came up with G-WIS—Gulf-wide Information System."

Figure 11.7
Spartina, a cordgrass common in coastal marshes, covered in oil from a spill in Dixon Bay, Louisiana, on January 28, 1995.

Photo by Dr. Terry McTigue, National Oceanic and Atmospheric Administration, National Ocean Service, Office of Response and Restoration.

Converting from paper to digital

Assembling the information for the database began with MMS working with NOAA to use that agency's ESI maps. The ESI maps, which were already widely used by NOAA and the oil companies, formed the basis for the G-WIS database. Although NOAA's ESI maps were produced from digital databases, most of the information was displayed on paper maps and each was specific to the particular areas it depicted. "In the past, each ESI database or atlas applied to only one state," Froomer said. "Although there was some degree of similarities from one state to another, each picked the peculiarities of the individual states."

When G-WIS began, Florida already had an oil spill response planning system—the Florida Marine Spill Analysis System (MSAS)—which was based on NOAA's ESI mapping, according to Henry Norris, a program administrator at Florida's Fish and Wildlife Research Institute. At the time, the states of Alabama, Louisiana, and Mississippi lacked the resources to build dedicated spill response databases and were dependent on paper ESI maps. G-WIS and MMS asked Florida and other states for updated data. "It was beneficial to us because then Alabama has the same system we do. The fish don't stop at Perdido Bay [on the Alabama-Florida border]," Norris said.

Source: National Oceanic and Atmospheric Administration.

Figure 11.8 Cleanup workers vacuum thick oil from the beach following a spill in Tampa Bay, Florida, August 1993.

Hunting and gathering data

Plans for creating G-WIS were discussed in numerous meetings. The participants were asked to either gather new data or to update existing data for the project, said Joanne Halls, an assistant professor of geography at the University of North Carolina at Wilmington. Halls was RPI's chief geographer during the G-WIS project.

"We had to work with a lot of the data providers to create metadata and document everything in order for it all to

Source: National Oceanic and Atmospheric Administration.

Figure 11.9 Heavy oil covers a mangrove habitat in Tampa Bay, Florida, August 1993, as the result of a spill.

Figure 11.10 A map based on Environmental Sensitivity Index data for Louisiana contained within the G-WIS datasets.

Source: Louisiana Gulf-Wide Information System, LDEQ Landsat Enhanced Thematic Mapper Pan-Sharpened Mosaic of Louisiana UTM15 NAD83, (2002) MrSID.

work together," Halls said. The RPI team members met with the states to make sure each understood exactly how its data was stored. "It's one thing to read the metadata, but you want to be very sure of how to properly convert data from one set of attributes to another, how you aggregate data together, or reclassify attributes," Halls said.

Advanced basemap information that went beyond the basic U.S. Geological Survey topographic maps often used as basemaps were also compiled by each state and added to the datasets. Halls and her team at RPI eventually acquired about thirty thematic datasets for each state. The data came in mostly as shapefiles or as coverages, but there was also some in AutoCAD drawing files. The process started with the ESI mapping concept that RPI had developed and added a more robust component for economic and socioeconomic data and improved transportation information.

Managing the process

To help manage and streamline the process, Halls assigned one RPI staff member to oversee the data gathering and integration process for each state. Compiling and integrating each state's data required constant communication with each agency and state.

"It wasn't easy, it took about a year, but it wasn't too bad because the staff I had at RPI was used to doing things like that," Halls said.

Before anything went to its final form, all data was sent back for approval. It was also printed out to help show how RPI converted the information to another format. With the conversions approved, everyone had the opportunity to view the data in the G-WIS format. When the data was edgematched, which had never been done before, some other states began to view mapping for wildlife and other environmental factors from new points of view. The edgematching process involves aligning features along an edge of one GIS data layer to the features of an adjoining layer. Viewing the final product provided an opportunity to see how all of the information was related. "For example, the fisheries people didn't know what the shellfish people were doing and they would see all that data compiled together and say 'oh wow, I didn't know what was happening over there,'" Halls said.

Source: Louisiana Gulf-Wide Information System.

Figure 11.11 Spilled oil forms a black band around tree trunks on a mangrove island in Tampa Bay, Florida, August 1993.

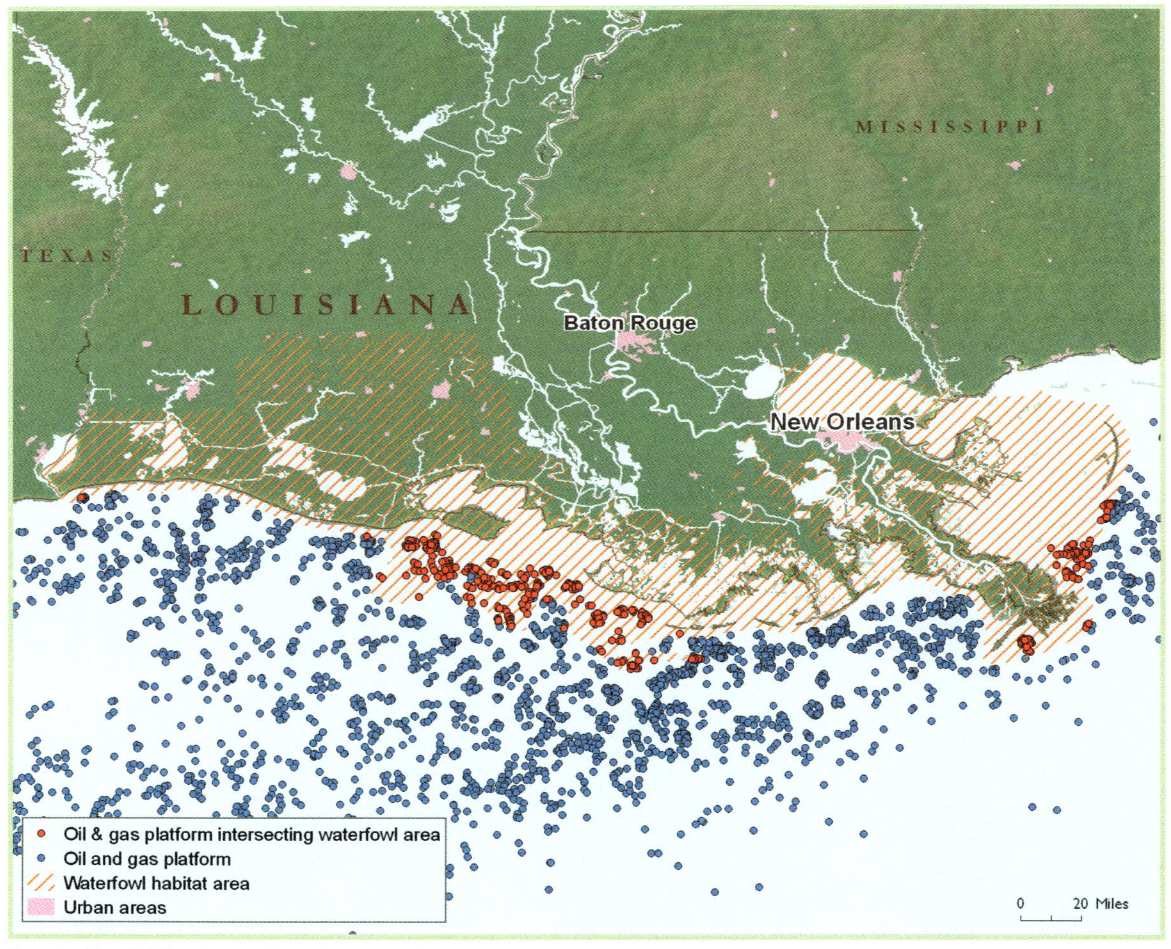

Source: Louisiana Gulf-Wide Information System, Environmental Sensitivity Index.

Figure 11.12 This map produced from G-WIS data shows the proximity of oil and gas platforms to critical waterfowl habitat areas in Louisiana.

The GIS that keeps on giving

It took three years, until 2000, to complete the G-WIS project. Although the result may not have met all the needs of all participants, those involved with the project agree it was beneficial and a major achievement in oil spill contingency planning and GIS database construction. The federal government and Gulf Coast states still use G-WIS in various forms. In Louisiana, which deals with more than six thousand reportable oil spills per year, G-WIS is used every time a significant spill requires investigation, according to Hanifen.

"It also comes in handy for permit issues, especially along the coast if there's some sort of construction project or pipeline project," Hanifen said. Halls added, "All of the states have taken the results of this project in different directions, but Florida has really raised the technology from the standpoint of emergency response. They take all of their computers out into the field and up into helicopters and they are very into rapid response using GIS."

Since G-WIS was incorporated in Florida's oil spill contingency planning database, it has morphed through several advances in both computer hardware and software, Norris said. When the G-WIS project began, Florida was running its ESI database within a Sun Solaris UNIX environment using ARC/INFO. G-WIS was originally conceived as a system that could run within the Microsoft Windows operating system, as its antecedents now do using the latest version of ArcGIS software. "What G-WIS did was build on the ESI paper [map] model and made it more robust and extended the information gulfwide," Norris said.

At the federal level, Froomer said G-WIS evolved into another system known as CORIS, the Coastal Offshore Resource Information System. CORIS, with its digital DNA from G-WIS, became the environmental component of the agency's corporate database.

Some features of G-WIS became very useful independent of oil spills. G-WIS introduced the concept of a study area, which defined the geographic limits of the area in which a study collected data. Study areas show either an absence of features because the features do not occur there or areas that show no features because the area hasn't been studied.

"It is very important to know this during an environmental review of an offshore platform or pipeline project," Froomer said. "If the project area shows no hard bottom features, for example, we know whether that's because there are no hard bottom features there or because there has been no survey there."

The MMS continued its working relationship with some of the partners in the project. About two years after the G-WIS project was completed, Florida updated its coastal resource database for its entire coastline. The state gave the updated data to the MMS, which in turn incorporated the information into CORIS at no expense to the MMS. In addition, Florida and the MMS will use the same data for environmental analyses and descriptions. This was one of the main selling points and promises of G-WIS. Observed Froomer, "If government and industry all used the same data structure and standards, we could save a fortune on data development. It was very gratifying to see that actually happening."

The significance of G-WIS

According to Norris, the creation of G-WIS gave the entire gulf, for the first time, a consistent oil spill response dataset. From a technical standpoint both Halls and Wayne agreed that G-WIS was a great leap forward. G-WIS has since become a template for other similar projects.

"It was a huge exercise in data management and we were tired at the end," Wayne said. "But it was quite satisfying because everybody had the same goal. The oil companies don't want to oil the marshes. Environmental damage is bad for everybody's bottom line."

Among the accomplishments of G-WIS Penland cited was validating the concept that a very large GIS could bring together data in a variety of formats from numerous sources and produce an authoritative information source that is accepted by the private sector and state and federal agencies. G-WIS was also created out of genuine collaboration between government and the oil industry, Froomer pointed out. "This was really a grassroots effort. It wasn't like it came from a policy out of the White House that the USGS was dictated to implement; it wasn't a trickle down," he said.

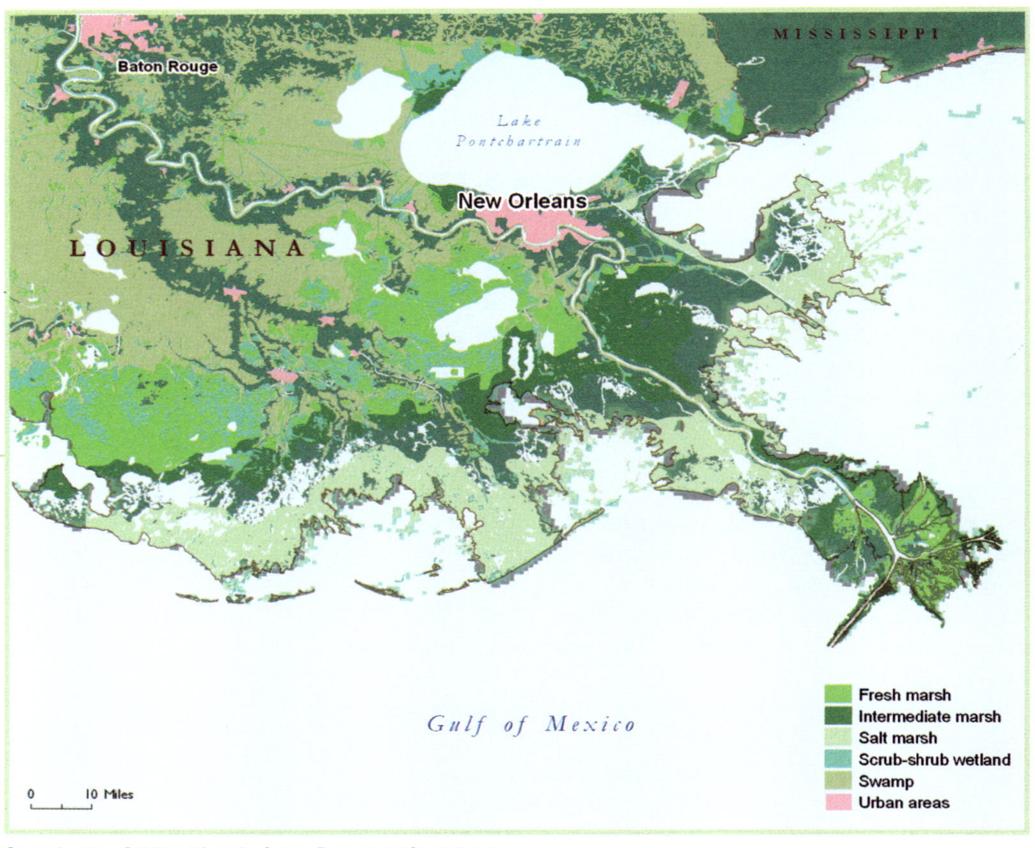

Figure 11.13 Various types of wetlands in southern Louisiana mapped using G-WIS data. Knowing where different types of wetlands exist is potentially useful for dealing with various types of oil spills.

Source: Louisiana Gulf-Wide Information System, Environmental Sensitivity Index.

NOAA goes fishing and G-WIS meets ELMR

The G-WIS project helped the National Oceanic and Atmospheric Administration (NOAA) devise comprehensive maps of the Gulf Coast's fish and invertebrate populations. One of the roles NOAA played in building G-WIS was to provide spatial information on the distribution and abundance of fishes and invertebrates in thirty-one estuaries along the Gulf Coast, according to Dr. Mark Monaco, chief of the NOAA Strategic Environmental Assessment Biogeographic Characterization Branch and one of the key participants in the project.

NOAA collected and synthesized data on the spatial distribution and numerical abundance for about forty-four species and organized the data into shapefiles. Most of the data originated with the five Gulf Coast states. NOAA organized it into a common spatial framework and developed ways to compare the relative abundance of similar fish and invertebrate types.

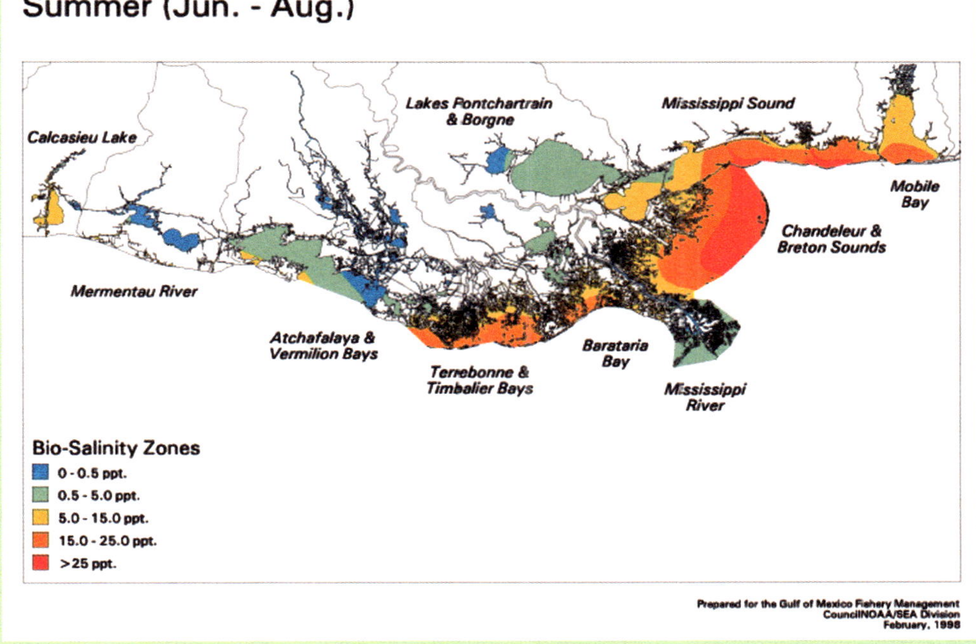

Source: Louisiana Gulf-Wide Information System.

Figure 11.14 Biosalinity zone map created from ELMR data for Louisiana, Alabama, and Mississippi.

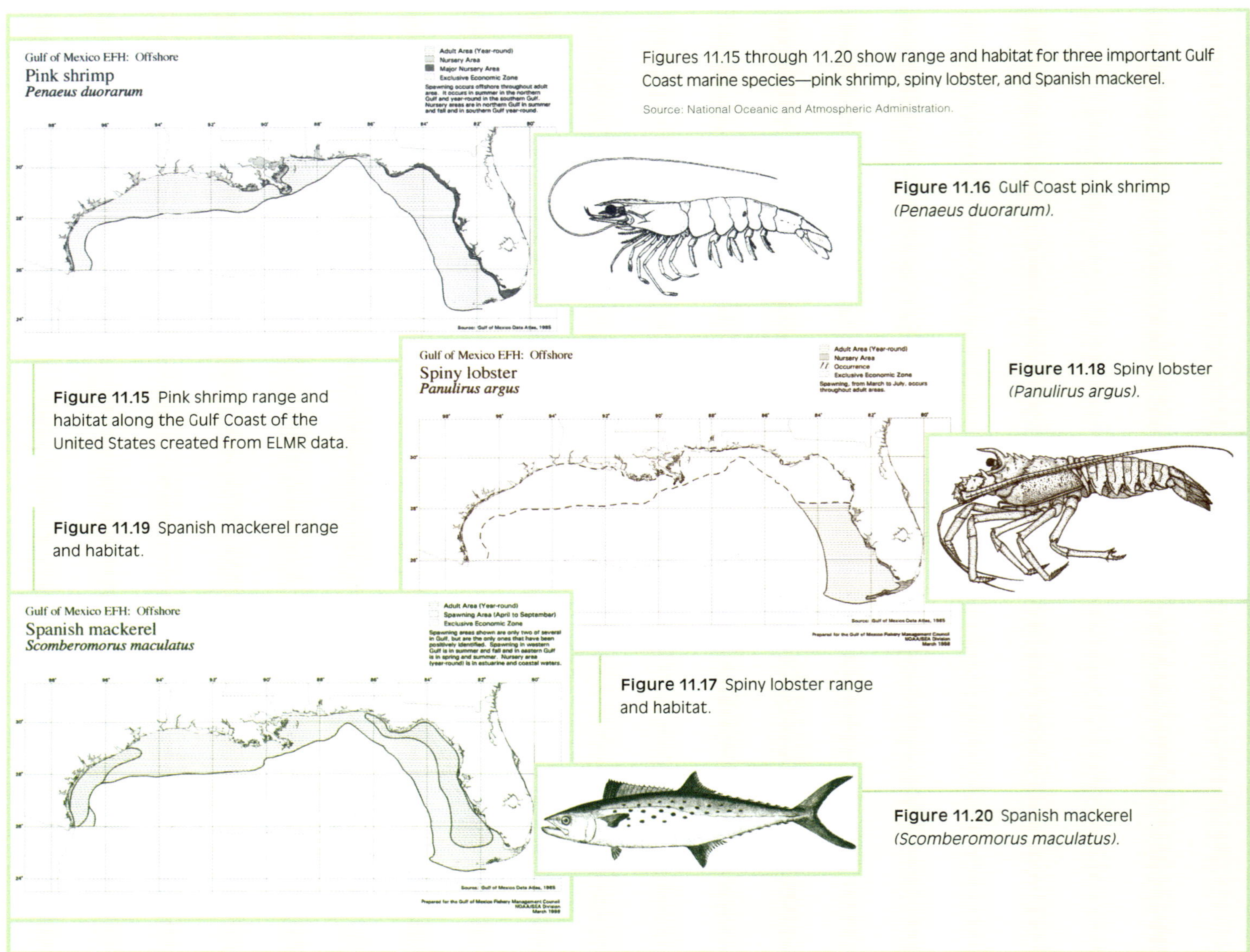

Figures 11.15 through 11.20 show range and habitat for three important Gulf Coast marine species—pink shrimp, spiny lobster, and Spanish mackerel.

Source: National Oceanic and Atmospheric Administration.

Figure 11.15 Pink shrimp range and habitat along the Gulf Coast of the United States created from ELMR data.

Figure 11.16 Gulf Coast pink shrimp *(Penaeus duorarum)*.

Figure 11.17 Spiny lobster range and habitat.

Figure 11.18 Spiny lobster *(Panulirus argus)*.

Figure 11.19 Spanish mackerel range and habitat.

Figure 11.20 Spanish mackerel *(Scomberomorus maculatus)*.

The NOAA database—Estuarine Living Marine Resources (ELMR)—was created with ArcView software. ELMR was originally a tabular database organized by salinity zones found in the gulf's estuarine environments and by the life stages of the animals involved. For each animal's life stage, NOAA came up with a relative abundance estimate.

NOAA worked with the states to update the tabular fisheries dataset and convert it into a spatial depiction of the salinity in each estuary. That dataset became the spatial framework for predicting the distribution of the animals based on their habitat requirements.

After the data tables and maps were created, state officials in a series of workshops reviewed the data to ensure its accuracy. "The GIS made that happen really easily," Monaco said. "Having the information displayed visually as maps made all of the difference in the world." He found G-WIS to be a significant project—considering the human and financial resources involved—that pulled all the data together, formulated it using ArcView software, and worked with the individual states to have it reviewed.

The end result was that ELMR became the estuarine fish and invertebrate database for G-WIS. "Ultimately it gave the database's user community . . . information on the distribution and abundance of these sensitive species in estuaries," Monaco said. NOAA was also able to use the G-WIS fisheries data as the basis for official habitat maps for a project that updated the ELMR tabular database and identified fish species that needed preservation and management. "The resources provided by MMS allowed NOAA to take datasets that were derived from the states and make [them] spatial," Monaco said.

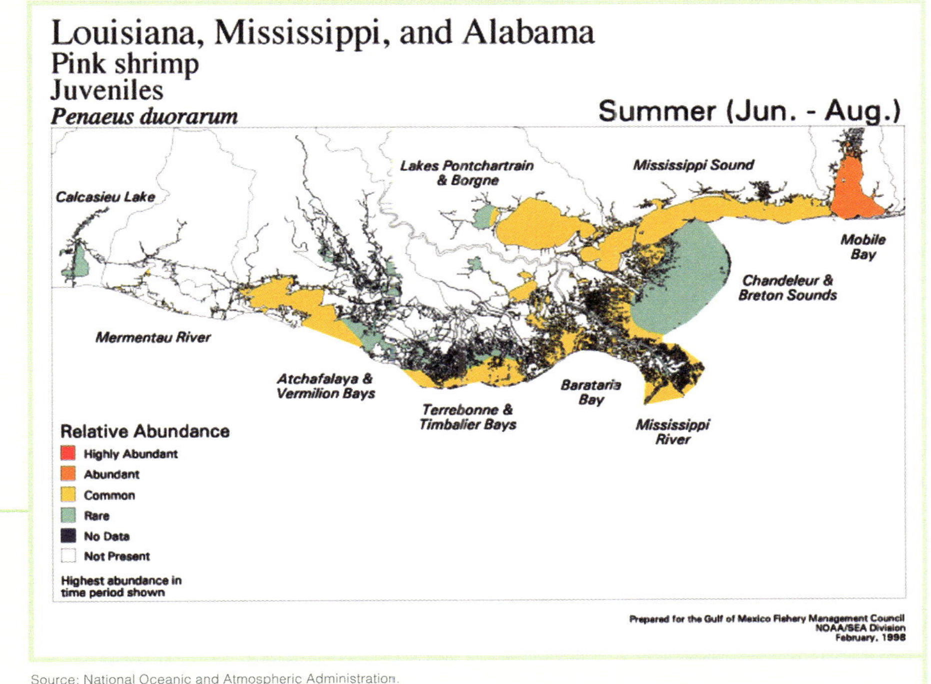

Source: National Oceanic and Atmospheric Administration.

Figure 11.21 Relative abundance map created from ELMR data for pink shrimp juveniles along the U.S. coastline of the Gulf of Mexico.

References

Energy Information Administration. Hurricane Katrina's impact on the U.S. oil and natural gas markets.
tonto.eia.doe.gov/oog/special/eia1_katrina_083005.html

Froomer, Norman. 1995. Evolution of an oil spill contingency planning GIS database. Paper presented at ESRI User Conference, 1995, May 22–25. San Diego, Calif.
www.gis.usu.edu/docs/protected/procs/esri/1995/to350/p347.html

Gulfbase.org, Research Database for Gulf of Mexico Research.
www.gulfbase.org

G-WIS Environmental Sensitivity Index datasets. Louisiana State University and the U.S. Minerals Management Service.
atlas.lsu.edu/central/esi/gwis-esi_faq.html

Public policy and regulation 12

Influencing environmental decisions in Maine

Augusta is Maine's state capital. A coalition of state agencies based in Augusta collaborated on Maine's first federated GIS.

Data provided by 2004 ESRI Data & Maps.

Sometimes the most difficult part of GIS isn't the technical details. Getting various organizations to work together can be the greater challenge. That was the case in Maine when GIS users collaborated on an innovative federated GIS system that has proven beneficial to a wide variety of players: Maine's legislators, state and federal regulators, businesses and commercial interests, and the public at large.

"The hardest part in the whole affair was getting the humans to sit down and agree on what they wanted," said Michael Smith, senior database analyst for the Maine Department of Environmental Protection (MEDEP). "The technology part was pretty straightforward."

Smith said creating a federated GIS involves assimilating data from numerous GIS systems, but still allowing each enterprise system to maintain its own GIS.

Why federate?

In Maine, GIS and IT professionals from several state agencies joined forces and convinced eight state agencies and one state program to create Maine's first federated GIS, similar to a federated database system. Several databases are integrated into one virtual database via a network to create a federated system. Users can access a federated database much as they would a central database, but these databases are not in the same location.

In a federated GIS, users can access a central GIS resource, which may consist of GIS data from numerous sources, applications, or services. A federated GIS differs from an enterprise GIS, which provides data, applications, and services to users throughout an organization. A federated GIS links separate enterprises that can interact with each other's GIS data. Each enterprise's total GIS capabilities are in turn available to an array of end users spread across multiple enterprise systems.

Maine's GIS evolution has its roots in the late 1980s when the state digitized data for early computer-based maps, Smith said. Enterprise GIS came along in the late 1990s, providing data, applications, and Web interfaces. "There are several agencies that have their own SDE servers," said David Kirouac, senior programmer analyst for the Maine Office of GIS (MEGIS), where he is responsible for many statewide GIS databases. "It's just kind of the way that things evolved."

But why federate when GIS users in the various state agencies already have enterprise systems? The answer is simple: a need for standards. The "federating" of GIS data is the next logical evolutionary step as enterprise systems grow increasingly interconnected.

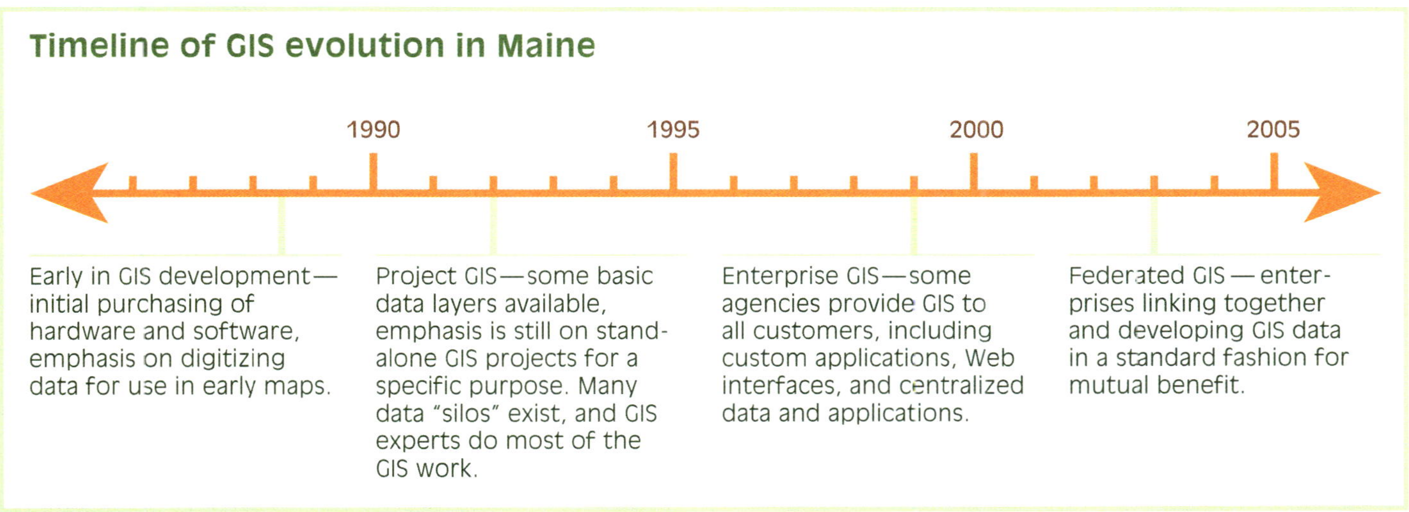

Adapted from Maine Department of Environmental Protection GIS.

Figure 12.1 How GIS evolved in Maine.

"It arose out of the need to concurrently edit the same hydro data in Maine," Smith said.

Problems arose when MEGIS would produce a hydrographic database that several state agencies and individual users would copy. Each would edit the information, so standardization was impossible. "You wound up with a thousand versions of the same dataset," Smith said. When the state's 1:24,000 hydrography dataset was due for revision, several state agencies seized the opportunity to standardize the data and reach agreement about how to maintain it.

Sharing data, working together

In 2003, three state agencies—the Maine Department of Inland Fisheries and Wildlife (MEIFW), MEGIS, and MEDEP—were updating various aspects of the state's hydrographic database. Kirouac was integrating the hydrographic data the U.S. Geological Survey and MEGIS created for the state. Once the base data was in, the three agencies would edit the hydro data. But first, the data from each state agency required merging. "We had to get the hydrography data to a place where what we had was a combination of the [three agencies'] data," Kirouac said. "We got to a common place with the data before we started any editing."

Getting the data to a common place was actually the easy part. Numerous meetings took place and, at first, other interested agencies attended. But it quickly became apparent that the three primary agencies involved with the data would edit the hydrographic database. Representatives from the three agencies determined each agency's jurisdiction and priority for the data. MEIFW wanted accurate maps of islands so it could track sea birds and seals. MEDEP's primary interest was in inland waterways.

Like any social system, a federated GIS requires rules and limits. A successful federated database depends on creating standards and controlling who can edit data. Each of the agencies involved in the federated GIS had to agree to common editing standards by using certain master data layers on a central server. Convincing the various agencies to move from a mindset of "mine" to "ours" was also important to the project's success, according to Smith. "It's doesn't do the state of Maine any good if we all have our own datasets," said Vicki Schmidt, a GIS environmental specialist with the Bureau of Land and Water Quality at MEDEP.

Figure 12.2 Federated GIS Web services in Maine's state government.

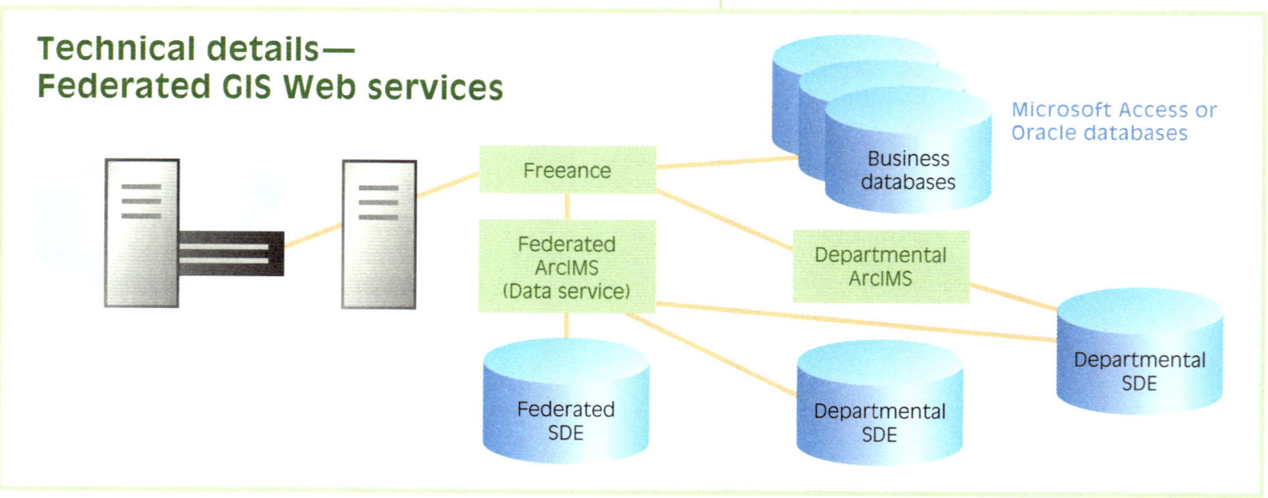

Adapted from Maine Department of Environmental Protection GIS.

Coming to terms

In the end, the agencies agreed that MEIFW would maintain the island boundaries while MEDEP and the Office of GIS would maintain the hydrographic boundaries. MEDEP also maintains a database of ponds, lakes, and rivers. An official memorandum of understanding outlined each agency's role in editing data and specified users who have the power to make edits. The memorandum also spelled out editing standards. The agencies provided feature-level metadata for the data layers or other information they provided.

Figure 12.3 Maine's Department of Environmental Protection maintains a database of ponds, lakes, and rivers, including the Kennebec River, which flows through Augusta.

Building the hydrography dataset began with the U.S. Geological Survey project that MEGIS was developing—the National Hydrography Database. Data produced by MEIFW on coastal islands was incorporated, and a layer showing lakes and other attributes from MEDEP's hydro data was used to complete the picture. The result was a federated GIS hydrography database capable of accepting simultaneous updates from three state agencies in a single SDE database. At the same time, the federated GIS is available on the Web to any state agency or the public. The ultimate purpose of Maine's federated GIS is to assist regulatory and public-policy decision making.

"With multiple agencies interested in the data, you need one way to edit the data, otherwise you wind up with multiple versions," said Schmidt. "You get maps showing up at legislative hearings that are mapped differently because someone has upgraded data or changed something."

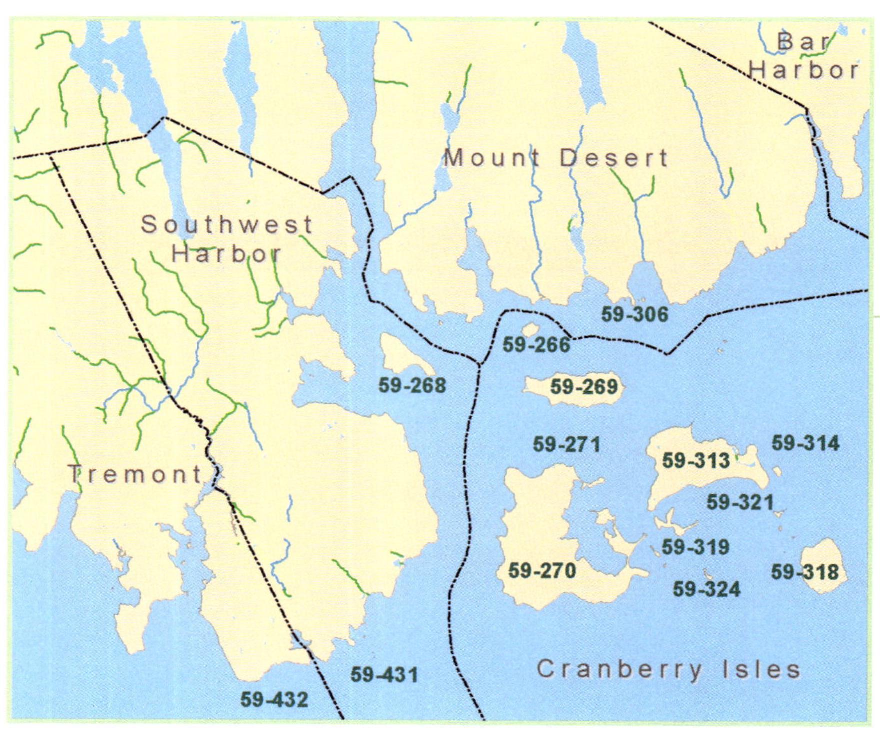

Source: Maine Department of Environmental Protection GIS.

Figure 12.4 Building the hydrography dataset in Maine began with work on the National Hydrography Database for the U.S. Geological Survey. Data on coastal islands was incorporated, and a layer showing lakes and other hydro data was used to complete the picture. The result was the ability to produce maps like this one.

Federated GIS succeeds

Schmidt said the federated GIS database had a positive affect on the public-policy process in Maine. Maps produced by the system helped to clearly convey information about water quality during legislative hearings. "We gave a visual aspect to the water quality classification law," Schmidt said. "Being able to represent that in a visual way certainly made things easier for people to understand."

Instead of publishing a three-inch-thick report full of tabular data, MEDEP was able to publish seven maps that showed in very graphic ways the changes in water quality around facilities such as paper mills and wastewater treatment plants over fifteen years. Other state agencies in Maine are creating federated GIS databases. For example, the Maine Drinking Water Program (MEDWP) and MEDEP simultaneously update maps of public water supplies and protection areas in a single SDE database. In the past, the two agencies kept separate databases. "We applied the same exact principle," Smith said. "We have one data layer and everyone updates through the Maine Office of GIS."

Maine was planning to turn two separate roads databases—one maintained by the Department of Transportation (MEDOT) and the other by MEGIS for the state's 911 emergency response system project—into a federated GIS database. MEGIS replaced static education GIS data with data actively updated by the Maine Department of Education. The education GIS data is available from the MEGIS SDE database but the attributes come directly from the education database and reflect changes to that database as they occur. Maine is also using ArcIMS software to serve up Internet mapping so that anybody with an Internet browser can create maps directly from the data in the federated database, Kirouac said.

Examples of federated GIS in Maine include the following:

- 1:24,000 National Hydrography Dataset—simultaneously updated in a single SDE database run by three state agencies
- Public Water Supplies and Protection areas—simultaneously updated in a single SDE database by two state agencies
- ArcIMS and Freeance Web services—shared Web architecture and templates multiple state agencies can use
- Numerous agency-specific data provided to other agencies via a single SDE database

Source: Maine Department of Environmental Protection GIS.

Serving the public with GIS data

ArcIMS software helps broaden access to the data because it allows members of the public who have no experience with GIS to view the data, including the hydrography dataset, Kirouac said. "We're here to try to serve the public, consultants, and paper companies, and anyone who wants to see the data and understand the resource," Schmidt said.

While the federated GIS has become a powerful new tool for state agencies and the public, it has also created new demands on the state government's GIS professionals. "I don't think it's made things easier from our standpoint as GIS people," Schmidt said. "If anything it means more work for GIS people. As more people get access to the data and want to do their own mapping projects, it just gets to be a nightmare from a management perspective."

Another example of federated GIS in Maine's state government is ArcIMS and Freeance software creating IMS Web sites. MEDEP purchased a Freeance license, but runs its Web sites on the MEGIS ArcIMS server to take advantage of that agency's IMS infrastructure. The Freeance license can also be used by other agencies because it is a per-server license.

"We hope that we can build upon what we have already," Kirouac said. Datasets recently included in the hydrography database include lake-depth measurements and dams. Other incremental additions and updates are continually in progress.

"Hopefully, one day we'll make this a full-blown model so people can do analysis," Kirouac said. "Each layer can correspond into this federated GIS." Other state agencies may jump on the federated GIS bandwagon. "We're looking at getting the Department of Marine Resources more involved because they have an interest in the coastal areas and do work that no one else does," Kirouac said. "This was something that we'd talked about for years. The people who were involved basically made it happen; that made it move forward."

Figure 12.5 Datasets recently included in Maine's federated hydrography database include depth measurements for lakes like this one in Baxter State Park, a 202,000-acre wilderness area.

References

Kirouac, David. 2004. GIS common editing environment. *Maine IS Technology* 7(5):6. www.maine.gov/newsletter/june2004/gis_common_editing_environment.htm

Kroot, Christopher, and Michael Smith. 2005. The evolution of federated GIS in Maine. Paper presented at ESRI User Conference, July 25–29. San Diego, Calif. gis.esri.com/library/userconf/proc05/abstracts/a1909.html

Smith, Michael. 2005. Evolution of federated GIS in Maine. PowerPoint presentation.

Glossary

brownfields Real property contaminated or potentially contaminated by a hazardous substance. The expansion, redevelopment, or reuse of brownfields may be complicated by the presence or potential presence of hazardous substances.

crosswalks Processes that involve converting classes in each original dataset to one common set of classes.

digital orthophoto quarter quadrangle (DOQQ) A computer-generated image of an aerial photograph that has removed image displacement caused by terrain relief and camera tilts. It combines the image characteristics of a photograph with the geometric qualities of a map.

edgematching A spatial adjustment process that aligns features along the edge of one layer to features of an adjoining layer. The layer with the least accurate features is adjusted, and the adjoining layer is used as the control.

federated GIS A system that allows users to access a centralized GIS resource, which may consist of GIS data from numerous sources, applications, or services.

gap analysis The scientific method of identifying and classifying components of biological diversity and determining which components are already in protected areas and which components are absent from or underrepresented in protected areas.

Genetic Algorithm for Rule-set Production (GARP) A genetic algorithm automates the use of environmental data collected through field surveys to produce distribution maps and models. This method has been largely applied to predicting the distribution of species of animals and plants, but can potentially predict any observable environmental entity.

geoprocessing GIS operations such as geographic feature overlay, coverage selection and analysis, topology processing, and data conversion.

Global Positioning System (GPS) The U.S. Department of Defense constellation of satellites that orbits the earth and transmits signals that allow a GPS receiver to calculate its own location. GPS technology is used in navigation, mapping, surveying, and other applications that require precise positioning.

ground truthing The process of conducting ground surveys to confirm details and features shown on maps, aerial photos, or satellite images.

image segmentation An iterative algorithm that aggregates pixels into groups of spectrally similar pixels known as regions.

land cover The classification of land according to the characteristic that best describes its physical surface; for example, pine forest, grassland, ice, water, or sand.

land use The classification of land according to what activities take place on it or how humans occupy it; for example, agricultural, industrial, residential, urban, rural, or commercial.

Landsat Earth-orbiting satellites developed by NASA and the U.S. Geological Survey that gather imagery for land-use inventory, geological and mineralogical exploration, crop and forestry assessment, and cartography. Landsat satellites have gathered images of the earth since 1972.

nearest-neighbor analysis A method of examining the distances between each point and the closest point to it.

model A set of rules and procedures for representing a phenomenon or predicting an outcome. Models can be created in ModelBuilder software and consist of one process or a sequence of processes connected together.

point-pattern analysis A way to study the location of events and reveal facts about the distribution of those locations. Point-pattern analysis can show whether the locations of events are randomly clustered or regularly distributed.

Ripley K analysis A type of point-pattern analysis that preserves distances at multiple scales and can quantify the intensity of patterns at multiple scales.

Tablet PC A handheld computer operated with a special pen instead of a keyboard or mouse.

t-tests Assesses whether the means of two groups are statistically different from each other.

Books from *ESRI Press*

Advanced Spatial Analysis: The CASA Book of GIS *1-58948-073-2*

ArcGIS and the Digital City: A Hands-on Approach for Local Government *1-58948-074-0*

ArcView GIS Means Business *1-879102-51-X*

A System for Survival: GIS and Sustainable Development *1-58948-052-X*

Beyond Maps: GIS and Decision Making in Local Government *1-879102-79-X*

Cartographica Extraordinaire: The Historical Map Transformed *1-58948-044-9*

Cartographies of Disease: Maps, Mapping, and Medicine *1-58948-120-8*

Children Map the World: Selections from the Barbara Petchenik Children's World Map Competition *1-58948-125-9*

Community Geography: GIS in Action *1-58948-023-6*

Community Geography: GIS in Action Teacher's Guide *1-58948-051-1*

Confronting Catastrophe: A GIS Handbook *1-58948-040-6*

Connecting Our World: GIS Web Services *1-58948-075-9*

Conservation Geography: Case Studies in GIS, Computer Mapping, and Activism *1-58948-024-4*

Designing Better Maps: A Guide for GIS Users *1-58948-089-9*

Designing Geodatabases: Case Studies in GIS Data Modeling *1-58948-021-X*

Disaster Response: GIS for Public Safety *1-879102-88-9*

Enterprise GIS for Energy Companies *1-879102-48-X*

Extending ArcView GIS (version 3.x edition) *1-879102-05-6*

Fun with GPS *1-58948-087-2*

Getting to Know ArcGIS Desktop, Second Edition Updated for ArcGIS 9 *1-58948-083-X*

Getting to Know ArcObjects: Programming ArcGIS with VBA *1-58948-018-X*

Getting to Know ArcView GIS (version 3.x edition) *1-879102-46-3*

GIS and Land Records: The ArcGIS Parcel Data Model *1-58948-077-5*

GIS for Everyone, Third Edition *1-58948-056-2*

GIS for Health Organizations *1-879102-65-X*

GIS for Landscape Architects *1-879102-64-1*

GIS for Water Management in Europe *1-58948-076-7*

GIS in Public Policy Using Geographic Information for More Effective Government *1-879102-66-8*

GIS in Schools *1-879102-85-4*

GIS in Telecommunications *1-879102-86-2*

GIS Means Business, Volume II *1-58948-033-3*

GIS Tutorial: Workbook for ArcView 9 *1-58948-127-5*

GIS, Spatial Analysis, and Modeling *1-58948-130-5*

GIS Worlds: Creating Spatial Data Infrastructures *1-58948-122-4*

Hydrologic and Hydraulic Modeling Support with Geographic Information Systems *1-879102-80-3*

Integrating GIS and the Global Positioning System *1-879102-81-1*

Making Community Connections: The Orton Family Foundation Community Mapping Program *1-58948-071-6*

Managing Natural Resources with GIS *1-879102-53-6*

Mapping Census 2000: The Geography of U.S Diversity *1-58948-014-7*

Mapping Our World: GIS Lessons for Educators, ArcView GIS 3.x Edition *1-58948-022-8*

Mapping Our World: GIS Lessons for Educators, ArcGIS Desktop Edition *1-58948-121-6*

Mapping the Future of America's National Parks: Stewardship through Geographic Information Systems *1-58948-080-5*

Mapping the News: Case Studies in GIS and Journalism *1-58948-072-4*

Marine Geography: GIS for the Oceans and Seas *1-58948-045-7*

Measuring Up: The Business Case for GIS *1-58948-088-0*

Modeling Our World: The ESRI Guide to Geodatabase Design *1-879102-62-5*

Past Time, Past Place: GIS for History *1-58948-032-5*

Planning Support Systems: Integrating Geographic Information Systems, Models, and Visualization Tools *1-58948-011-2*

Remote Sensing for GIS Managers *1-58948-081-3*

Continued on next page

When ordering, please mention book title and ISBN (number that follows each title)

Books from ESRI Press (continued)

Salton Sea Atlas *1-58948-043-0*

Spatial Portals: Gateways to Geographic Information *1-58948-131-3*

The ESRI Guide to GIS Analysis, Volume 1: Geographic Patterns and Relationships *1-879102-06-4*

The ESRI Guide to GIS Analysis, Volume 2: Spatial Measurements and Statistics *1-58948-116-X*

Think Globally, Act Regionally: GIS and Data Visualization for Social Science and Public Policy Research *1-58948-124-0*

Thinking About GIS: Geographic Information System Planning for Managers (paperback edition) *1-58948-119-4*

Transportation GIS *1-879102-47-1*

Undersea with GIS *1-58948-016-3*

Unlocking the Census with GIS *1-58948-113-5*

Zeroing In: Geographic Information Systems at Work in the Community *1-879102-50-1*

Forthcoming titles from ESRI Press

A to Z GIS: An Illustrated Dictionary of Geographic Information Systems *1-58948-140-2*
Charting the Unknown: How Computer Mapping at Harvard Became GIS *1-58948-118-6*
GIS for the Urban Environment *1-58948-082-1*
Mapping Global Cities: GIS Methods in Urban Analysis *1-58948-143-7*

Ask for ESRI Press titles at your local bookstore or order by calling 1-800-447-9778. You can also shop online at www.esri.com/esripress. Outside the United States, contact your local ESRI distributor.

ESRI Press titles are distributed to the trade by the following:

In North America, South America, Asia, and Australia:
Independent Publishers Group (IPG)
Telephone (United States): 1-800-888-4741 • Telephone (international): 312-337-0747
E-mail: frontdesk@ipgbook.com

In the United Kingdom, Europe, and the Middle East:
Transatlantic Publishers Group Ltd.
Telephone: 44 20 7373 2515 • Fax: 44 20 7244 1018 • E-mail: richard@tpgltd.co.uk

ESRI Press • 380 New York Street • Redlands, California 92373-8100 • www.esri.com/esripress